Hilderic Friend

The Ministry of Flowers

Hilderic Friend

The Ministry of Flowers

ISBN/EAN: 9783337306700

Printed in Europe, USA, Canada, Australia, Japan

Cover: Foto ©berggeist007 / pixelio.de

More available books at **www.hansebooks.com**

THE
MINISTRY OF FLOWERS

BEING

𝕾ome thoughts respecting 𝕷ife, suggested by
the 𝕭ook of 𝕹ature

. BY THE

REV. HILDERIC FRIEND, F.L.S.

AUTHOR OF
"FLOWERS AND FLOWER LORE," ETC., ETC.

LONDON
SWAN SONNENSCHEIN & CO.
PATERNOSTER SQUARE
1885

𝔓𝔢𝔯𝔱𝔥:

S. COWAN AND CO., STRATHMORE PRINTING WORKS.

I Dedicate

THIS LITTLE BOOK TO MY CHILDREN

IN RECOGNITION OF THEIR LOVE

OF THE BEAUTIFUL IN

NATURE.

Perth:

S. COWAN AND CO., STRATHMORE PRINTING WORKS.

I Dedicate

THIS LITTLE BOOK TO MY CHILDREN

IN RECOGNITION OF THEIR LOVE

OF THE BEAUTIFUL IN

NATURE.

TABLE OF CONTENTS.

BOOK I.

Respecting Human Life.

BOOK II.

Respecting the Evils of Life.

b

BOOK III.

Respecting the Virtues of Life.

BOOK IV.

Respecting Various Features of Life.

PREFACE.

FLOWERS! How many a sermon have they preached, and how many a text have they supplied! The Saviour says—" Consider the Lilies how they grow, they toil not, they spin not ; and yet I say unto you, that Solomon in all his glory was not arrayed like one of these" (Luke xii. 27). And centuries before He uttered the parable of the Mustard-seed or of the Fig-tree, Jotham had spoken as follows : " The trees went forth on a time to anoint a king over them; and they said unto the olive-tree, Reign thou over us. But the olive-tree said unto them, should I leave my fatness, wherewith by me they honour God and man, and go to be promoted over the trees? And the trees said unto the fig-tree, Come thou, and reign over us. But the fig-tree said unto them, Should I forsake my sweetness, and my good fruit, and go to be promoted over the trees? Then said the trees unto the vine, Come thou, and reign over us. And the vine said unto them, Should I leave my wine, which cheereth God and man, and go to be promoted over the trees? Then said all the trees unto the bramble, Come thou, and reign over us. And the bramble said unto the trees, If in truth ye anoint me king over you, then come and put your trust in my shadow; and if not, let fire come out of the bramble, and devour the cedars of Lebanon " (Judges ix. 8-15).

Every true lover of flowers must have been struck by the fact that these gems of earth are capable of imparting large stores of knowledge, if only the eyes are open to see, and the heart to receive them. We cannot understand the narrowness of mind which exists in some good people who love the sight of the flowers as they grow in copse and meadow, but cannot tolerate their use for charitable or religious purposes. They allow wretched imitations to adorn their houses in the shape of wall-paper, and chimney ornament ; they allow the sculptor to carve representations on the pillars of the sanctuary, and they do not protest against their symbolic use on stained glass or in the border of some painted panel, bearing the Paternoster or the Creed ; yet they will not admit the pure, innocent, fragrant, and surpassingly lovely originals, as they come direct from their Maker's hands into His sanctuary. But this unreasonable prejudice is gradually giving way, and at stated times and

seasons in many places among several of the various
sections of the Christian Church, flower services,
and harvest thanksgivings, and other similar
gatherings for praise and adoration are being held,
and made attractive by the use of fruits and
flowers. Under these circumstances the thought
has frequently occurred—could not some further
interest be thrown into these gatherings by means
of the Ministry of Flowers? The following pages
supply some hints and suggestions which, it is
hoped, will enable many to answer that question
in the affirmative. The volume has been divided
into books, in order that the various lessons which
the flowers teach us might be the better arranged
in some definite order, and those matters which
seemed most appropriate for the use of such as
intend to give addresses on the subject are placed
first. For many years the study of flowers and
flower-lore has been the congenial pastime with
which I have relieved myself when wearied with

other labours, and I may therefore claim to have a somewhat full acquaintance with the subject from personal investigation. I owe much to those who have written on kindred topics, and would here especially refer to the suggestive works of Dr. Taylor, Dr. Cooke, and others.

May this little work both lead to the further study of the works of nature, and assist the reader to love with greater intensity the great Creator of all that is beautiful around us !

HILDERIC FRIEND.

BOOK I.

⁓⁓⁓

The Ministry of Flowers

RESPECTING HUMAN LIFE,

⁓⁓⁓

" Nature has a spiritual as well as a material side."
Rev. Hugh Macmillan.

A

BOOK I.

———

𝔥𝔲𝔪𝔞𝔫 𝔏𝔦𝔣𝔢.

LIFE is the first great concern of man. He is ever considering what he shall do in order to ensure his further existence, and all that he possesses will he give for life. It is not therefore of unimportance that we should inquire —what do the flowers say to · us respecting this vital matter? And though it may not be to many a very cheery thought with which to open a book on so bright a theme, yet we are constrained to remark that the most important lesson which we learn from flowers relates to

THE BREVITY OF HUMAN LIFE.

We will ponder well this solemn matter at the very outset, for if we can dwell on it to our profit we shall then be able to turn to other views of life with a light and joyous heart. Let us hear first the messages which echo through the ages, and find confirmation in every land. "In the morning men are like grass which groweth up. In the morning it flourisheth, and groweth up ; in the evening it is cut down, and withereth " (Psalm xc. 5-6). "As for man, his days are as grass : as a flower of the field so he flourisheth. For the wind passeth over it, and it is gone ; and the place thereof shall know it no more " (Psalm ciii. 15-16). Christ speaks of " the grass of the field, which to-day is, and to-morrow is cast into the oven " (Matthew vi. 30). The Eastern oven is very different from our own. I have seen them in China made exactly after the same fashion as those employed in Palestine, consisting of " a covered earthen vessel, wider at the bottom than at the top, wherein bread [or small cakes] was baked, by putting hot embers round it, which produce a more equable heat than in the regular oven." Here, too, as in the Holy Land, you will see the poor women on the hill sides cutting down the rank grass and flowers for burning in these jar-shaped ovens, for "the wild flowers which form part of the meadow-growth, are counted as belonging to the grass, and are cut down with it. Cut grass," adds Dean Alford, "which soon withers from the heat, is still used in the East for firing." The prophet tell us (Isaiah xl. 6), "The voice said, Cry. And he said, What shall I cry? All flesh is grass, and all the goodliness thereof is as the flower of the field : The grass withereth, the flower fadeth ; because the Spirit of the Lord bloweth upon it : surely the people is grass. The grass withereth, the flower fadeth ; but the word of our God shall stand for ever." These words St. Peter quotes and endorses also (i. Peter i. 24). But long before the time of the prophet we find the patriarch Job exclaiming : "Man that is born of a woman is of few days, and full of trouble. He cometh forth like a flower, and is cut down " (Job xiv. 1-2). The Singer of Israel in his Prayer of the Afflicted complains (Psalm cii. 11), " My days are like a shadow that declineth ; and I am withered like grass." Nor is this the portion of the poor merely, for St. James reminds us that the rich shall fade away as the flower of the grass. . "For the sun is no sooner risen with a burning heat, but it withereth the grass, and the flower thereof falleth, and the grace of the fashion of it perisheth ; so also shall the rich man fade away in his ways " (James i.

11). In the East one cannot fail to observe how rapidly the choicest vegetation, and most brilliant flowers, droop and die. The intense heat of a tropical sun soon deprives the earth of its moisture, and carries off the life-giving and health-sustaining qualities hidden in the soil, and nature all around wears a parched and sombre aspect. Nor is it to the sun alone that the early death of the flower is due. The worm enters the plant as it did the gourd which Jonah grieved over, and even the stately Cocoa-Nut Palm bows its head and ceases to yield its fruit when attacked by these destructive vermin. Death in many instances is retarded by the timely detection of the evil, and nimble lads are employed to dig the worms out from the heart of the affected tree, and so prolong its valuable life.

There are probably few of us who take any interest at all in flowers—and who does not love them?—who have not observed that certain kinds exist only for a very brief period. There are perennials whose life may be counted by years; there are annuals, whose work is done usually in the course of a single summer; and there are others which spring up, blossom and die again in a very short time. Then we are aware that though in many instances the plant itself may live for a considerable time, the flowers are very short-enduring. Thus we have all heard of the Changeable Rose (*Hibiscus mutabilis*), so called

MALLOW.   [*face p.* 6.

on account of the fact that though the flower, when first it opens, is white, it soon changes to rose-colour, and then to purple. In the West Indies all these changes take place in the course of a single day, but when the plant is brought to these climes a week is required for the process. This flower is a native of the East Indies, from which parts the French carried it to their settlements in the West Indies, and to it they have given the characteristic name of *Fleur d'un heure.* Another species of Hibiscus is the Venice Mallow, which is a native of Italy and Austria, bears a purple and yellow flower, and was formerly known in our English gardens as Good Night at Noon. The quaint old herbalist, Gerarde, remarks that "it openeth itselfe about eight of the clocke, and shutteth up againe at noone, about twelve a'clock when it hath received the beams of the sun, for two or three houres, whereon it should seeme to rejoice to look, and for whose departure, being then upon the point of declension, it seemes to grieve, and so shuts up the flowers that were open, and never opens them againe; whereupon it might more properly be called *Malva horaria,* or the "Mallow of an Houre." We have all observed how the flowers of the Convolvulus have drooped at the close of the day, and failed to raise their bell-shaped heads again.

BUT if the flowers speak in solemn tones of the brevity of human life, they also supply us with many interesting illustrations of

LIFE IN ITS VIGOUR AND GLORY.

Some flowers possess the property of retaining their hue and life-like appearance long after they have been gathered, on which account they have been called Everlasting Flowers. But of these we must not now speak, as we wish to show rather that there are many plants and trees which are capable of withstanding the storms of centuries, till eventually they become the very picture of majesty, and so tell us of longevity. The reader will be aware that if we have no historical means of telling when a tree was planted, we can in many instances ascertain its age by counting the rings formed by the annual production of new wood. Now there are some trees whose age when

they died could not be much less than that of Adam or
Methuselah, and there are even instances of Yew trees
planted in England which have endured for centuries the
heat and cold, the sunshine and storm of our changeful
clime. Trees of such endurance will in many instances
be of enormous size, and as the cedar, the glory of
Lebanon, is frequently employed in Scripture as a
representative of great men, so the majestic Oak or stately
Sequoia may perform the same kind office. The prophet
in one or two instances couples together trees of various
kinds in order to represent persons of different degrees of
influence or position in life, as when he says—" The
glory of Lebanon shall come unto thee, the fir tree, the
pine tree, and the box together, to beautify the place of
my sanctuary ; and I will make the place of my feet
glorious " (Isaiah lx. 13). The Ash has been known to
reach to great proportions. What was supposed to be
the greatest tree of the kind in England was recently
felled at Aber, in North Wales, under instructions from
Lord Penrhyn, and we are told of one which used to
exist at Logierat which was sixty feet in height, and forty
feet in girth at a yard from the ground. Among
Oak trees, that known as the Cowthorpe measured at
its base seventy-eight feet, and was supposed to be some
sixteen centuries old. The Salcey Oak was computed to
have lived through one millennium and a half, and was

forty-six feet around its trunk ; while a Chestnut tree at Tortsworth can be traced back over seven hundred years, and is believed to be a thousand years old. It measured some fifty feet or more around the stem. The oaks of Sherwood Forest are still famous. Near Worksop, Notts, there formerly stood a tree, "which, in respect both to its own dignity and the dignity of its situation, deserves honourable mention. In point of grandeur few trees equalled it. It overspread a space of 90 feet, from the extremities of its opposite boughs. These dimensions will produce an area capable, on mathematical calculation, of covering a squadron of 235 horse. The dignity of its station was equal to the dignity of the tree itself. It stood on a point where Yorkshire, Nottinghamshire, and Derbyshire unite, and spread its shade over a portion of each. From the honourable station of thus fixing the boundaries of three large counties, it was equally respected through the domains of them all, and was known far and wide by the honourable distinction of the 'Shire Oak,' by which appellation it was marked among cities, towns, and rivers in all the larger maps of England." It is well known that in former times many boundaries up and down the country were fixed by means of trees, which, in virtue of their size and well-defined position, would readily serve as landmarks, and so come to be objects of interest, and even of respect.

Everyone who pays a visit to the Dukeries, or roams among the former haunts of brave old Robin Hood, is expected to go to the "Parliament Oak," which, in the form of a crumbling fragment, stands on the brow of a gentle hill in the neighbourhood of the old hunting-seat of King John. This famous tree is supposed to be upwards of a thousand years old, and though its decayed form can only be kept together by means of props and supports it had once a majestic and beautiful form. Tradition says that on one occasion, while Edward I. and a noble and princely band of followers were hunting in Sherwood Forest, a messenger arrived in haste, bearing the unwelcome news that Wales was in open revolt, and must be at once brought into subjection again, or endless mischief would ensue. Promptly the monarch gathered around him his knights and followers, and the proud oak afforded them shelter as they discussed in open council the subject of the hour, and under its majestic form the knights, with true and loyal feeling, declared themselves willing and determined to support their sovereign in his attempts to suppress the insurrection, and stamp out the evil. One oak in this forest was some years ago made into a kind of triumphal arch, by being cut through the stem in such a way that a carriage could be driven through it; and this reminds us of a famous tree in Normandy, which, a century and a half ago, was converted into a place of worship, and went by the name of the Chêne Chapelle.   Its trunk was at that time hollow, and its head partly decayed.   Having been paved and provided with a roof, this unique building was divided into two apartments, the lower of which, by the liberality of the Abbé du Détroit, was fitted up as a chapel, while the upper formed a chamber for the officiating priest !   This Oak is about equal in size to the Greendale Oak, to which I have just referred, the arch through which is several inches higher than the entrance to Westminster Abbey, known as the Poets' Postern, under which men pass on horseback, and through which carriages are said to have been driven.   So famous is this oak in the neighbood of the forest that several towns and villages have an inn bearing the painted sign of the Greendale Oak.

Reference may here be made to one or two other remarkable trees, whose age and size may be regarded as fit emblems of the glory and dignity of human life.   Seven centuries ago the Great Chestnut of Tamworth (or Trotsworth) was referred to in writings, which are still in existence, as a signal tree ; and if in the year 1135 A.D. it already merited the title of " The Great Chestnut," it is not difficult to believe that in its youth it was contemporary with the Saxon Egbert, and has witnessed the fortunes and failures of a thousand years.   The famous Chestnut of the Hundred Horse, which formerly existed on Mount

Etna, is an historical marvel. One writer asserts that its ruined trunk gave proof of having measured a hundred and sixty feet ; and the fact that the peasants built a house within it, in which they had an oven for drying chestnuts, is the strongest evidence of the truth of this calculation. Its name of Castagno di Cento Cavalli is said to have been derived from the circumstance that Jean of Arragon and her attendants, who numbered one hundred nobility on horseback, obtained shelter from it during a storm which overtook them on Etna. It must suffice that I here mention the Baobabs merely as further illustrations of the enormous size and age attained by trees in some parts of the globe. The Sequoias of California are world renowned. The wood is very light, and the trunks tower away above the head of the traveller, till they seem to be lost in the clouds. Imagine a tree which at 400 years of age is still but in its youth, and which does not begin to tremble with age before it has stood a thousand years ! These famous trees grow on the slopes of the mountains, and are kept supplied with the best of food in the shape of minute particles of fertilizing rocks, pulverised by the weather, and the decayed vegetation of ages. And what we should suppose Noah or Enoch to have been—rugged and majestic in their latter-day glory—such are these trees. " There is no symmetry in his top, or delicacy and grace in his outline ; he has battled and struggled with the storm for too many centuries to preserve an artistic appearance. He looks the giant of the forest, broadrooted and strong-limbed, rough and weather-beaten, but defying snow and frost and hurricane for thousands of years, and sheltering bird and beast and cattle beneath his grand shadow." One tree known as the " Old Maid," in the Calaveras grove, was about 1370 years of age, and had a delicate waist of upwards of a hundred feet in circumference. It took three men the long period of twenty-five days to cut her down, which they did by boring the trunk with augers. From three to four hundred feet is the average height of these botanical wonders, and who can look at them or think of them without exclaiming, "How wonderful are Thy works, O Lord." We leave the reader to consult such works as relate to our woodlands and forest trees for further particulars, and he will experience no difficulty in finding many records of trees equally remarkable for size with those we have briefly enumerated.

I T will scarcely be from such trees as I have been describing that we shall find illustrations of our next subject, which relates to

THE STRUGGLE FOR LIFE.

These seem to have as thoroughly established themselves in their present position by the right of long possession as it is possible for them to do, and it will be a hardy race of warriors that will be able to bear away the Palm from the gigantic Sequoia, or the robust Oak. But if we descend from the lofty heights to which our musings have led us, and now pursue the study of the minuter forms of life, we shall find that there are the most palpable indications on every hand that a mighty struggle for life has been, and still is, going on. And, indeed, how can it be otherwise? The earth is limited in its extent, and

cannot possibly produce beyond a given number of
plants, that number being regulated by the size and
qualities of the various individuals. In a plot of ground
twelve feet square you might grow some fifty cabbage
plants, and a thousand weeds would thrive apace. But
plant a Yew tree or two, or set a graceful Fir in the
centre, and these would soon exterminate the rabble
which sought a resting place at their feet. Here in very
truth the battle is to the strong, and the race to the swift,
and strength and fleetness are constantly being called into
action. I shall have by-and-bye to dwell on progression
and retrogression in life, so shall not here refer to the
recent discussions of various botanists and evolutionists
in respect to the inherent properties possessed by plants,
but shall at once proceed to give some facts in illustration
of our present subject—the struggle for existence. We
find it in the world of nature everywhere. Never was it
fiercer or more intense than now in the business and
political worlds, never, perhaps, more marked than now
in the religious world, and it is quite certain that the
struggle is as strong to-day among the lower animals and
plants as ever it was. The world is just composed of an
innumerable army of parasites. If you catch a shark
you find a lesser fish living on the monster's skin ! If you
kill a pigeon or a fowl you will find it alive with insects,
some of which are the hosts of other minuter creatures !

If you pluck a nettle you will find a thousand fungi (*Æcidium*) feeding on its leaves, and the Oak is laden with mosses and lichens and insects of a hundred different kinds ! Does not all this indicate that life is earnest, and that in order to sustain it, everything must adapt itself as best it may to the circumstances in which it is found? This adaptation to circumstances is very marked in the flower world, and abundance of fruitful lessons may be learned by observing the methods employed. We must not overlook the fact that the flower and tender plant have as many enemies as man himself, and that they have to hold themselves in readiness to meet these opposing forces. Sometimes the enemy of a slender flower is to be found in a stronger rival which thoughtlessly thrusts its broad leaf or coarse stem right over the tender bud or blossom of its humbler sister. Sometimes it is a tree or hedgerow, a bank, rock, river, or lake which comes in the way, and throws the plant into new circumstances which necessitate one of two things—ignominious death, or altered form. Sometimes as a protection against animals the plant defends itself, for on its power to do this does its continued existence depend. Have you not observed that the Holly usually bears leaves with strong spines or prickles all about its lower part, while these unpleasant appendages are wanting or are greatly softened and modified on the upper boughs where they run no

B

risk of being browsed? Its struggles to exist when sur-
rounded by quadruped foes led to this device, and similar
devices can actually be traced historically in the development of many plants which have had to stand on the
defensive. Look for example at the Nettle. Who has
not felt the influence of its sting as he has thrust his hand
into the hedge for a nest, or has been searching in the
bank for some coveted flower? Now the sting of the Nettle
is evidently a device by means of which the plant is able to
keep at bay some of its greatest enemies, and so secure
an existence in the world during the great race of life.
The Hedgehog and the Nettle would go well together, and
we might then couple together the Porcupine and the
Great Nettle found in Sumatra, the effects of whose
poison are such that weeks of suffering have resulted to
the unwary botanist or native who has been so unfortunate
as to grasp the ill-natured thing. Many people defend
themselves in a similar way, and we do not wonder that
they are allowed to live on while many gentler and
worthier spirits succumb to the influences under which
they live, and fall in the battle of life.

Along with the prickly Holly and the stinging Nettle,
we may rank the various kinds of Thorns, Briars, and
Thistles with which we meet in this, and other lands.
The golden Furze, the rich and showy Hawthorn, the
Blackthorn, and several other rosaceous flowers, including

the Dog-rose and Bramble, the Raspberry and Goose-berry, the Restbarrow, and a hundred other plants, adopt a similar method of defending life and property from the assaults of animal foes.

In many instances the fruit, upon the germination of which the continued existence of the family depends, is encased in a horny, woody, or prickly covering. Every boy who has picked up the chestnuts which fall in autumn from the trees remembers how the finely pointed needles from the outer case have entered his fingers and defied all attempts at dislodgment. The leathern inner cover-ing of the chestnut, horse-chestnut and acorn are similar protections, and being of a rich brown colour they may easily lie undetected among the autumn foliage and decay-ing vegetation until they have found a congenial spot for growth and then put forth their plumule and roots. On the other hand we find many kinds of fruit, especially those which we know as stone-fruit, encased in a soft pulpy covering which is particularly agreeable to birds and animals, and leads to their being devoured and carried away to a favourable spot where the uninjured kernel is ejected and assisted in its growth by its congenial sur-roundings. It is thus that the Sloe, Cherry, Yew, Haw-thorn, and many other plants are able to establish them-selves in new places and so keep up their family name. Certainly in the sharp struggle for life the lesson thus afforded is a very pleasing one, and we have reason to believe that many have been able to fare well in days that have gone by adopting a policy similar to this.

One of the most ingenious devices with which we are acquainted, and one which is familiar to every one who has ever been in the country, is that which we find most largely employed by the flowers of the Composites, such as the Dandelion, Groundsel, and Thistle. When the seeds are ripe we find them on a fine dry day taken up by the wind and borne to a great distance, there to be dropped in the hope that they will find a spot in which to grow. The seeds are sustained in their flight by a very ingenious arrangement of the Pappus as it is called, which spreads out in the form of an umbrella and so catches the passing breeze and glides merrily away. In the Tragopogon this arrangement is of so striking a character that the plant has in consequence acquired the name of Goat's-beard. But I need do no more than refer the reader to his own experience, and he will instantly call to mind the Dandelion "clock," and the way in which the hour of the day was ascertained by blowing at the seeds till they yielded to the force of a mouth-made hurricane and betook themselves to some distant spot. In some plants we find a still further attempt to secure a footing in the earth, and one which must lead us to ad-mire the wisdom of the Being who created the plants, and endued them with these qualities, or at any rate placed within them the inherent potentialities, which, as occasion serves, they have developed or evolved. Everyone knows the curious Horse-tail, as the Equisetum is called. It looks very like a tail standing on the stout end and bristling all over with stiff hairs. This plant has a peculiar method of producing fruit and propagating itself, and in some points it comes near the fern in its generative apparatus. What is usually spoken of as its fructifying organs form a cone or spike at the extremity of certain of the branches, the barren and fertile forms being easily distinguished by this means. This cone consists of a cluster of shield-like disks, each of which bears a circle of spore-cases which open longitudinally in order to set free the spores or rudimentary buds, by means of which the plant is multiplied. Now each of these spores has attached to it, a pair of elastic filaments, that are originally formed as spiral fibres on the interior of the wall of the primary cell within which the spore is generated, and are set free by its rupture ; these are at first coiled up closely around the spore, but on the slightest application of moisture they suddenly extend themselves, so that when they fall to the ground and are brought into contact with any moist body, they throw out their grappling-hooks and fix themselves upon the spot. The study of these spores under the microscope is intensely interesting, and if a friend breathes upon them while the examination is going on, their elastic motion may be watched and studied to perfection. Without these elaters the spores would stand as poor a chance of being able to exist in the midst of so much competition as would the seed of the Dandelion or Thistle without the valuable aid of its pappus. That life is a struggle may easily be seen by the fact that not only have many plants become dwarfish and insignificant as the ages have rolled by, giving place to others of more vigorous type and daring habit as we shall see in the chapter on retrogression, but also from the circumstance that many plants have given up the contest altogether, and now exist only in the herbarium of some museum, or in the stone cabinets of the earth in the form of fossils. It has fared with plants as with animals, and the same holds true respecting families and races of men, while some have by dint of perseverance, hard work, and a good deal of scheming and contriving managed to hold their own and even to make rapid advances, others have had to give up their territory inch by inch and foot by foot, until at last they have been fairly beaten and could try no longer.

I have mentioned but two or three methods by which the end of life is gained ; many others might be enumerated. Some plants, for example, continue to exist because they are possessed of some virulent poison or juice, or are so sticky that their enemies are kept by them at arm's length. I well remember some years ago finding a beautiful spurge in a Devonshire lane and plucking its fruit as I walked along in order that I might taste the quality of the capers. For some hours my tongue and throat suffered from the irritation, so powerful was the property it contained.

ESPITE the earnest efforts which are made by the flowers and plants to hold their own, we see sad evidences of

RETROGRESSION IN LIFE

on every hand, and to this matter we will now give some attention. Everyone will remember the stir which was made a few years ago by the publication of certain works bearing on the subject of Evolution, and the Survival of the Fittest. Now the very word "survival" implies the idea of failure, and if the fittest only survive, the weakest must of necessity gradually be overcome and crowded out. The doctrine of Evolution in its widest application includes degeneration or retrogression as well as development or progression. Dr. Taylor well says that the vegetable kingdom is not less fruitful than the animal in proofs of this law "that, while the main mass of living organisms have throughout geological time advanced to higher ground, some have stood still, or merely 'marked time,' and others have gradually lost ground and dropped out of the ranks. Some of the latter have honestly struggled to keep up the step, and have found themselves unable; others have done so only by diplomacy and cunning, or by the development of qualities even more sinister" ("Sagacity and Morality of Plants," p. 207). Look, for example, at the Horsetails (*Equisetum*). As known to us to-day this group of curious plants is of diminutive stature and circumference, yet, if we trace them back to earlier times by means of their fossil representatives, we find that at a certain period they measured from six inches to a foot in diameter; while they gradually assume in the older strata an arboreus form, and may be regarded as actual trees, just in the same way as our so-called tree moss, which grows some six inches high, is to be seen on the hills around Canton (China) reaching almost to the dignity and dimensions of a shrub. Some of the so-called moss-ferns of geological times are of immense size, their modern representatives being mere dwarfs and pigmies beside the earlier race of giants. One writer has summed up our knowledge of fossil plants in the following words, which, though written half a century ago, are still found to be true: "First, in the oldest strata in which land plants occur, Ferns are met with in the greatest abundance. Secondly, their absolute numbers and relative proportions become wonderfully diminished in the superior formations, until, in the later series, they are

comparatively scarce. Thirdly, not only is their numerical
strength astonishingly lessened, but they are still more
remarkably reduced in size." (" Outlines of Botany,"
p. 341.) If I appeal to Mr. Carruthers for his testimony
in this matter, his words will be worthy the warmest
reception, for we have few higher authorities on the
subject of fossil plants. He tells us (*Contemporary
Review*, February, 1877, p. 401) : " The later Palæozoic
rocks abound in plant remains. The first evidence of
land-plants is found in the Devonian rocks; and here, at
their appearance, the three principal groups of the
vascular cryptogams appear together in highly differen-
tiated forms. All of these—Ferns, Horsetails, and
Lycopods—possess the same essential structure and
organization as their living representatives, and in all the
subordinate points in which they differ from them it is in
the possession of characters indicative of *higher organization*, whether we look at the vegetative or the reproductive
organs, than are found in existing forms. Thus among
Ferns there is lost a remarkable group with a fundamentally different type of structure, which was contem-
poraneous in the Palæozoic ages with the types of Ferns
that have been represented all through the epochs, and
are now abundant on the globe. The Equisetaceæ had
a large number of generic groups ; their stems were
arborescent, the leaves large, and the fruit comes pro.

tected by special scales ; but the spores were similar in
size and form to those of the living Horsetails, and were
even furnished with hygrometric elaters [see p. 21].
The Lycopods were also huge trees, and were represented
by several generic groups. The stem structure, while
fundamentally agreeing, like that of the aborescent Equisetaceæ, with the stems of the living forms, was more complex, being suited to their aborescent habit." There
were giants in those days, and we are forcibly reminded
of the Bible records from which we learn that in former
times men attained a much greater stature and age than
they do at present. As animals and plants have increased
in number they have gradually diminished in size ; and
while in some instances this diminution has only resulted
in the further concentration and development of existing
powers, in others it has been followed by decay and
retrogression. These results are in some instances trace-
able to what are called natural causes. We all know
what an influence climate has upon plants and animals.
There must have been a time, judging from the fossil
plants which we find in England and Europe, when the
climate of this island and continent was very different
from what it is to-day. Now if plants which luxuriate in
heat and moisture, and can only thrive to perfection in a
natural hot-bed, are gradually placed in a colder and less
congenial atmosphere and soil, they will develop much

less rapidly, and will eventually become dwarfed and inconspicuous. In this way some of the facts before us are to be accounted for. We see this in the case of the Ferns, for while those which grow in our own land are usually small, we see them attaining the dignity and importance of trees in other lands where the climate is more congenial. One of the most interesting signs of retrogression may be found in the fact that whereas some of our commonest flowers once boasted the full number of floral appendages in the form of sepals and petals, they have been compelled to give up some of these luxuries, just as a gentleman who has " seen better days " will have to part with his butler, coachman, and horses, when reverses come upon him. I one day picked up a specimen of the field Speedwell (*Veronica agrestis*), and was interested by observing that, instead of four petals—the ordinary number found in this plant—there were but three; and since the one that was missing is usually paler than the other three, the loss was compensated for by the manner in which two of the trio here found were shaded off to a lighter colour near their margins, which occupied the position of the missing petal ! Here was a plant which had formerly been keeping four servants, three with a light blue livery, and one with a suit of grey. But the latter had been dispensed with, and his livery divided among two of the remaining number, one only of

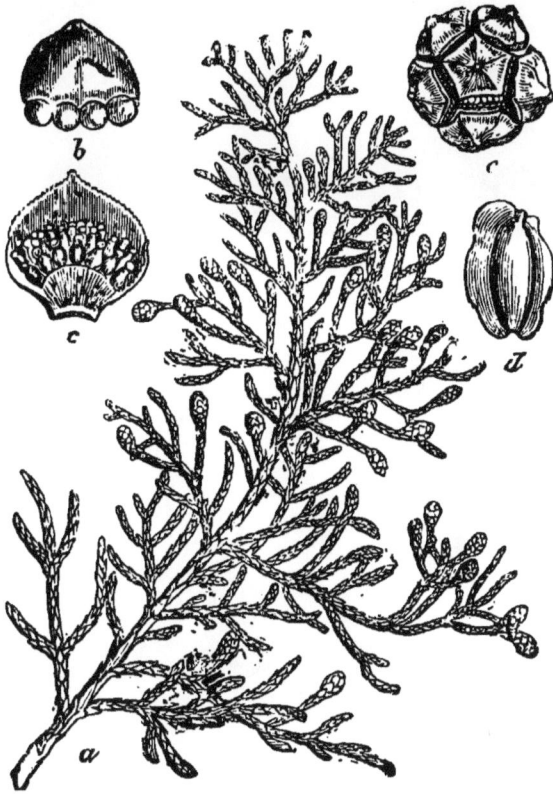

LYCOPODIUM. [*face p.* 28.

the retinue—the upper servant—still retaining its original position and dignity !

In China, one constantly sees exposed for sale, on the little street stalls, a dirty-looking nut, about an inch long, which is very largely bought by the poor people, who take off the outer covering, and amuse themselves with chewing the kernel.* This is the curious Ground-nut (*Arachis hypogœa*), or Earth-pea, quite different, however, from our English Ground-nut (*Bunium flexuosum*), or Pig-nut, as we perhaps more frequently call it. The English plant bears an edible root, or tuber, which may be found at the depth of two or three inches under-ground, and is much prized by school children. But in China, the nut is the fruit. In this curious plant, the uppermost flowers are barren and sterile, those alone bearing fruit which grow on the trailing or procumbent branches near the ground. After the blossoms have fallen, the stalks or peduncles elongate, so that, when the pods enlarge, the stem on which they grow is long enough to allow of their being buried beneath the surface of the soil. This plant is a native of America, the West Indies, and Africa, but is largely cultivated elsewhere, for the sake of the oil which its nuts contain. Let the foregoing facts be borne in mind for a moment, while we look at a similar pheno-

---

* Called "Monkey-nuts" in England, where they are now sold in the same way, and for similar purposes.

menon in our English Flora, and seek for the key to the mystery. Along by the sea-side near St. Leonards, I recently found a plant which has long been looked upon with wonder. It is one of the Clover family (*Trifolium subterraneum*), as the leaves at once tell us. Now, the Latin name *subterraneum* is exactly equivalent to the Greek *hypogæa*, both meaning "under-ground;" and this designation has been applied to the plant from the fact that, as the three delicate white flowers, of which the head usually consists, begin to wither, they turn downwards to the ground. From the end of the peduncles, there are produced a number of white fibres, arranged in the form of a star. These are, in reality, abortive calyces, "like small waxen hands with fingers out-spread," as the Rev. H. Wood aptly says. Within these calyces are the seed pods, or legumes, enclosed as if in a cage, to prevent their being lost; and, as the peduncles bend toward the ground, the pods finally work themselves under the soil, where they lie buried till the time for germination arrives. So small is this plant that you may easily walk over it without knowing of its existence; and it is only by searching for it that you will be able to find the slender wax-like blossoms, and observe the curious history of the plant. Now, why should these plants be so curiously constructed? The fact that they are nearly related to showy and imposing papilionaceous flowers—some of which arrive at the dignity of trees—leads us to infer that they have been gradually receding from their former more exalted position, and have by degrees developed this happy method of retaining their hold upon the soil, when they would otherwise have been exterminated altogether. This idea is further substantiated by the facts that we are daily becoming more familiar with, respecting the devices adopted by a great number of humble plants, whose relatives proclaim them to be fallen members of noble families. Thus, in Madagascar, we find an under-ground pea (*Cryptolobus*), the seeds of which are boiled and eaten while still unripe. These peas are produced in pods, which are buried beneath the soil, just as those of the Arachis and Trifolium are. It is not a great step from the Clover to the Wood-sorrel (*Oxalis*), for both have trifoliate leaves ; and the honour of being identified with the Irish Shamrock is equally shared between them. The Wood-sorrel, or Cuckoo-sorrel, is a plant too well known to need any description. In our own country, it blossoms about the time when the cuckoo arrives, and as it prefers the shady wood and hedge-bank—where, as yet, few insect visitors are to be found—it stands but a poor chance of being fertilized by their agency. Now, if it failed in this, it must either die out and become extinct, or provide for the contingency. This interesting plant has acquired more than one method of keeping up its stock, and that

which specially concerns us here relates to the production
of what are called cleistogamic flowers. These may be
looked upon as ordinary flowers, which have been arrested
in their growth, and which contain within themselves all
the properties necessary for self-fertilization, so that, without opening, they can ensure the production of seeds,
which shall produce in their turn a new progeny of plants.
You will find these cleistogamic flowers also in the Violet,
and, a little later on, you will observe seed-vessels, loaded
with polished seeds, ripening at the ends of peduncles,
which deposit their burden on the soil, close to the parent
plant, and so perform their task. The question arises—
" Why, if these plants have properly-organized and fully-
developed flowers—as we know the Violet and Wood-
sorrel to possess—why should it be necessary for them to
grow fruit on the sly ? " The study of a large number of
different plants—their habits, manner, and time of flowering, and other similar details—enables us to conclude that,
where plants have had to adopt these devices, it is evidence
of retrogression. If a gentleman of your acquaintance,
who once kept a large establishment, and carried on a
brisk trade, now comes to you, and solicits your orders
for a pound of tea or a cheese, you begin to think things
are going wrong. You never knew him do so before.
His well-stocked and smartly-arranged windows sufficed
for advertisements. But now, grocers have multiplied,

the customers are divided, and yet other methods than mere shop-window advertisement must be adopted. So with the flowers. If they cannot keep pace with the keen competition for the visits of insects, they must gain their living in other ways which are consistent with the dignity of their family and position. It has been truly remarked that, "in this bitter fight with poverty, there is a touching episode savouring of humanity. As much of the old show is kept up as the plant can possibly afford, and there are few species which do not bear ordinary flowers, as if nothing were the matter ; whilst the dwarfed and aborted cleistogamic flowers are hidden out of sight at the bases of the clustering leaves, as though the plant were anxious they should not be seen. The best face possible is put on the case, and often not without good results, for the occasional crossing which the conspicuous flowers of these plants get enables the seeds to gain back some of their old vigour, or to stay off the evil days of extinction, in which pure cleistogamism might end. The conspicuous flowers are not borne every year by some plants : they cannot afford such a luxury ; and one or two known kinds bear flowers which are of no good whatever, for they are never found fertile ; so, in their case, we must regard the habit as a survival, or as an indisposition to give up the old floral life and rank." (" Sagacity and Morality of Plants," p. 223.) We may take it for granted that, as a

general rule, all those flowers which depend on self-fertilization are incapable of progression ; and when, from
cross-fertilization by the agency of insects they revert to
the older method of producing seeds, they are already on
the decline.

But I will now call attention to another fact in reference to this subject, and we shall see that not only does
geology prove that some plants have been gradually de-
generating, nor are we taught it alone by the study of
those flowers which have contracted the habit of producing cleistogamic flowers. We find another evidence in
the colours of the petals or blossoms which some
flowers produce. In my garden I last year planted a
number of very choice Pansies, and was delighted with
their gay appearance. When I put them in the border,
I took care so to arrange them that the darker colours
should alternate with the lighter. But as we were sitting
at the window one day a friend remarked—"Why, you
have put all your light Pansies on one side of the bed,
why did you not mix them ?" I was startled, for there
was no denying the fact that all the blossoms produced
by the Pansies to the northward, where the sun seldom
shone, had lost their rich colour, and had dwindled down
to the most ordinary and unattractive flowers. It was
not long before I found that every plant in the bed had
degenerated in the same way, and the flowers were now

exactly like the common wild Pansy, except that the petals were larger ! Sometimes this degradation or retrogression is due to the soil, either it is too poor or too calcareous, or possesses too much clay. Thus you will find Hepaticas revert from pink, lilac, or blue, to white on a given soil, and take up the original colour if retransplanted to the original soil. This will account in large measure for the different tints we find in our wild, and especially sweet Violets. I know a part of Oxfordshire in which sweet Violets, ranging from the richest blue to the most delicate white, through a curious lilac and rose, may be found in successive strata so to speak, each variety being peculiar to a given soil and formation. Near Hurstmonceux Castle I found last year a number of most charming white Bluebells ! Indeed such freaks are very common among flowers whose normal colour is blue. So too we find red flowers becoming white, and many a child has gone home with a perfect prize when he has found a white Herb-Robert (*Geranium robertianum*). Of course the fact is more striking as you study the changes to which the highest orders and most thoroughly developed plants are liable. You do not notice the fall of a poor man as you do that of a wealthy land owner, whose carriages have daily passed your door, and whose bright-liveried servants have always been conspicuous objects. So we do not as readily observe the retrogression of such small flowers as Speed-wells, and Violets, and Pimpernels, as of Dahlias and Asters and Chrysanthemums. As everyone is familiar with the Primrose, both in its wild and cultivated states, I may ask the reader to observe for himself the way in which the petals of this flower change their colour. A plant which you have placed in a certain soil and situation will bear flowers of waxy whiteness, but its twin sister will produce blossoms of a dingy lilac.

Many other facts might be adduced, and the few illu-strations here given might be indefinitely multiplied, but the object I wish to keep in view is rather the stimulation of thought and observation than the supply of all the matter which is at hand. Those who keep their eyes open, and make notes of such facts as come under their own notice as they ramble by the hedgerows or spend an hour in their flower garden, will soon find that they have food for meditation and profitable reflection, which will not easily be exhausted, and no more instructive lesson will be learned than that on which we have here been dwelling, that in life everything has its reverses, and that men, as well as the lower animals and plants, are liable to such changes as their skill and industry will prove insufficient, without a higher power, to meet.

E may now turn to a more congenial topic, and see what the flowers teach us respecting

PROGRESS IN LIFE.

In the study of this subject we are on firm ground ; for everyone is aware that the flowers and plants we see around us are in many instances the proud and dignified descendants of less showy and serviceable ancestors. The Apple, Plum, and Nut can be traced back to the Crab, the Sloe, and the Hazel, which were formerly the sole representatives of many of the wonderful varieties of fruit we are now able to produce in such profusion. The gardener, by cultivation, cross-fertilization, forcing, dwarfing, pruning, budding, grafting, and a number of other skilful devices, has been able to work wonders in the world of nature, and what has been done for our cultivated fruits and flowers has also been done in many instances in a wild state by the flowers themselves. The scientific floriculturist follows out the laws which he finds already existing, and which he has seen in operation among the highest kinds of plants, and like causes give like results. The fertilization of a flower by its own pollen tends to keep it in its present state, or may prove detrimental, hence flowers are in many instances cross-fertilized. This fact the gardener is familiar with, and acting upon the laws he finds in operation around him, he places the pollen of one choice flower upon the pistil of another, and so secures a strong succession of plants which in many instances differ in colour, form, and quality from their ancestors. Now progress in life may be illustrated in a variety of ways. We might refer to the colours of the flowers, the size and variety of the fruits, their rich quality, their hardiness or delicacy, the endurance of the plants, their increased size, the number of their descendants, their advancing adaptation to meet their own requirements and to secure the best attention of the insects which wait upon them, and other similar topics ; and from these we should learn how we might act in order to further the great ends of life.

Let us begin with the colours of flowers, since these supply us a ready means of showing how development and progress are brought about. The simplest colours are white and yellow. You will be˙ struck by this fact as you begin in the spring time to collect plants for your herbarium, or as you go out to gather the first nosegay of vernal blossoms. Among the white flowers, some of which will perhaps show pink tips to their petals, you will find the Daisy, Chickweed, Whitlow Grass (*Draba verna*), Snowdrop, Violet, Dead Nettle, Rue-leaved Saxifrage (*Saxifraga tridactylites*), Christmas Rose (*Helleborus niger*), Shepherd's Purse, and other early flowers. Your nosegay will include such yellow flowers as the Lesser Celandine (*Ranunculus ficaria*), Daffodil, Dandelion, Crocus, to which you will gradually add the Yellow Thistles, Goatsbeard, Groundsel and its congeners, Yellow Flag, Hawkweeds, and a whole host of others. Now in many instances it will be found either that these flowers are self-fertilized, or are propagated by means of their roots (as in the Iris and Daisy, Crocus and Daffodil, Snowdrop and Violet), or, if visited by insects at all, only enjoy the favourable notice of tiny creatures which can venture out earlier in the season than the butterfly or bee. This is not by any means the case with all white and yellow flowers; for the evening Campion, evening Primrose, and other similar flowers, blossom at night and are visited by moths which are led direct to the blossoms needing fertilization by means of their light colour and grateful perfume. Still we may say that, as a rule, yellow and white are the simplest colours and may be found associated with plants of the lowest orders. By this we do not mean plants of tiny growth, for, curiously enough, some of the floral pigmies of our own and other lands stand very high in the ranks and give evidence of careful attention to the laws of progress. Ascending in the scale of colours we come to pink, red, blue, and purple, as well as to orange, and such as are parti-coloured and variegated. In the case of several families of plants, we find that the various representatives bear various uniforms, and generally the larger flowers bear the brightest and most attractive colours. If you study these flowers in groups, you will frequently notice that small flowers, low colour forms, and self-fertilization go together; while large plants, bright blossoms, and fertilization by insects go hand in hand. Take the English wild Geraniums by way of illustration, and from the tiny and unsavoury Herb-Robert, to the gaudy and elaborately devised meadow Geranium (*G. pratense*), what a grand advance do we observe ! If colour is the standard, what progress has been made in the development of flowers from the simple, old-fashioned, scarlet Geranium of the cottage window to the new and "improved" varieties yearly advertised in the catalogues of our leading florists. The moral is—cultivate your powers; rub your intellects against your neighbours'; take some of their pollen to fertilize the flowers of your imagination, and give them some of yours for a similar purpose. A man who evolves out of his own little self, his own stock, lives upon it, dies with it, and neither borrows nor lends, gives nor takes, may be great in his own eyes, but he becomes an object of disregard and even scorn to his neighbours.

Even to-day there are many people who have not learned to appreciate the potency of evolution. They have somehow been misled into the idea that to embrace the doctrine of evolution means shelving the Deity. It is a profound mistake. Rightly viewed, the doctrine of evolution does not detract from the power of God or His divinity (Romans i. 20), in the least, it rather leads us to wonder at, and admire the wisdom of the Being who could store up within the creatures of His hand such marvellous potentialities. If we say that the inhabitants of the Friendly Islands have become changed from savages and brutes to men and Christians, that they have evolved some noble qualities and are now a credit to themselves and the world, we do not for a moment detract from the power of religion and the agency of Christianity thereby. All has not been put within them. They were touched by a power that unlocked the secret springs and sources of greatness which were dormant in their natures, and which they obtained in common with ourselves by the infusion of the spiritual and divine when man was created. The possibility, the potentiality was there, now it has been developed and unfolded.

Thus is it with the lower animals, thus is it too with the flowers and plants. Some flowers which have never ranked higher than weeds till their secret powers have been discovered by some venturesome insect, have, through the agency of that creature, been enabled to rise to a noble rank in the aristocracy of plants, and from bearing flowers of the simplest nature have risen to beautify and adorn the place in which they grow. Just as the savage Briton has sped along the path of progress through the infusion of new life from without and the development of latent powers from within, till he takes foremost rank in all that concerns the well-being of the world, so from a wild weed to a rich and useful plant, the same progressive strides may be observed. Our tiny Groundsel is despised, save by the lad who keeps a pet bird, and we think it a disgrace to our gardens; but its sister the Cineraria, a highy developed form which has come from the Canary Isles, is placed on our tables to make them attractive when a dinner party is to be given, or a tasteful effect must be produced. From the one to the other what a wide gulf exists so far as the untutored eye can see! But why should so many colours be displayed, and why should such variety exist? Selfish people have but one answer—"These colour exist in order to gratify the sense of the beautiful in man; in order to please our eyes and make the country beautiful and attractive." Such is the reply which till recent times has given entire satisfaction, but we get impatient of old notions, and we soon ascertain that many of the most lovely forms of life existed long before man ever strutted in his pride across the land exclaiming, "I am monarch of all I survey." We find still that the tiniest insects, the smallest forms of animal and plant life, the most inconspicuous parasites, and a thousand other things on which the eye of man never rests, are of the most exquisite beauty and finish, some of them being carved and chased, painted and enamelled with a skill that puts us to the blush. Why all this display? The answer is—"Some wise purpose has been kept in view, some useful end is to be answered." And this thought has given to the study of botany a new life. Whereas it was once sufficient if we collected so many plants and tacked to them a Latin name, date and locality, now we find it necessary, if we wish to become botanists, to know why a flower has blue or orange petals rather than white or green, and why the stamens, pistils, pollen and other details in this flower differ so entirely in their shape, arrangement, number, and position from those in its neighbour. And when a question of such interest has been raised science at once comes to the rescue. The idea is suggested that possibly certain colours are specially attractive to certain insects, and that the shape of this flower or that is pecu-

liarly suited to the visits of a butterfly, a moth or a bee.
Then the secret gradually comes out, and we learn that
after all man's pleasure is quite a secondary consideration; that the first question relates to the plant itself;
that the fertilizing agent comes in for the next considera¯
tion, after which man may place himself where he pleases.
It is a simple thing to prove that flowers have progressed
in the art of colouring their petals. Some of the old
names of our commonest flowers have now become perfect misnomers by reason of this fact. What a change,
for example, has come over the Chrysanthemum since the
day when it was first christened the Golden Flower! And it
is certain that the Rose has undergone like changes, even
if we admit that its name is not originally derived from
a word meaning red. Homer speaks in such a way of
the dawn as to lead us irresistibly to the conclusion that
in his day the Rose was of a definite colour, and that
conclusion is confirmed by the many epithets in use
drawn from this distinguishing colour of the rose. We
hear of the rosy-fingered morn, rosy cheeks, a roseate
hue, but since these epithets became established we have
become familiar with the presence of roses red, and
roses white, of yellow and purple and pink and black
roses, and roses with variegated and parti-coloured petals.
If we take the family of plants to which the queen of
flowers has given her name, if we study the Rosaceæ, we

shall find that the large group of flowers collected under this designation includes a great variety of plants ranging from weeds on the one hand to trees on the other. In a few instances, as for example the Lady's Mantle, we find proofs of retrogression, then in the cinque-foils we find the normal type with yellow flowers ; a little higher up and we meet with such as have white flowers with pink tips; then lastly to the shrubs and trees with petals of crimson and rose-colour. The lower we go the nearer we get to the original type, till we even get as far beneath it in the case of some plants as we do above it in others. It is as though you take the family of a certain duke, and while on the one hand you find that certain members of the family rose higher and higher till they occupied the throne, others sunk so low in the scale of life as to be disowned by their richer relatives, and we only recognize them now by their name, or peculiar type of countenance, or by the study of their pedigree. The Poppy family *(Papaveraceæ)* is another interesting group, for while we find the gaudy scarlet petals of our Corn Poppy, by the side of the bluish white of the Opium Poppy, and the yellow of the Welsh variety, we find that the Horned Poppy and Celandine (*Chelidonium majus*, quite distinct from the lesser Celandine), together with the Eschscholtzia which we grow in our gardens, alike retain the original colour. The less attractive flowers have remained where they were, the others have made progress, and are now to be found establishing themselves everywhere. But this will suffice with reference to colour, and must also be taken as a sample of the kind of evidence we find in other directions. It would have been interesting to dwell on progression as witnessed in the fruits which different plants have borne, and this is a field abounding in facts if we take a wider scope and look at those tree and plants which are natives of other lands. The Orange, Lemon, Pumelo and other members of the Citron family have had a long and interesting history, and they have not been slow to avail themselves of the opportunities which have been placed within their reach. Here we find fruits as small as a marble, such as the Wang-pi (*Cookia*) of China, and others as large as a man's head. All are now more or less fit for food, but the curious variety of Citrons known as the Fingered Lemon or Buddha's hand is one of the survivors perhaps from earlier times, its fruit being used merely as an ornament, or for presenting at the heathen shrines. Some families of plants contain members which have branched off in different directions. Here is the fruit of the Soap-Berry tree (*Sapindus*) which I brought from China, where it is employed for making a lather as a substitute for soap. This rosary of 108 beads is made of the round, hard seeds which exactly resemble peas in size and shape, and which, in former times, were mounted in gold and silver and worn as buttons by the wealthy folk of our native land; the Spaniards still using them for beads and buttons. But while this plant has been content to produce its seeds in an inedible pulp, its sister (*S. esculentus*), the Pittombera of Brazil, produces fruit which is wholesome and eatable; that of yet another variety (*S. Mukorossi*) being innocuous, but very bitter. Now, if we take up the first cousins of the Sapindus we shall find that they have been more generous, and since they have favoured man with their fruits they have been rewarded by being placed in the fruit garden, and assigned an honourable place among the fruits of the East. The Lichi and Longan (the *Dimocarpus* and *Nephelium* of botanists) are very highly prized in China, and the fruit is often dried and sent to this country, where it is greatly enjoyed by some who are fortunate enough to be able to purchase it. It is said that entire trees are conveyed by water from Quang-tung (the province of Canton) to Peking, in order that the Emperor of China may enjoy this delicacy fresh and in its perfection. This is not merely a proof that his majesty is served with regal magnificence, but indicates the progress which the fruit has made, and the honourable place to which it has attained. Flowers and fruits, plants and trees, are still growing, developing, and progressing.

They are utilizing the private stores which have been
secreted within the cells and tissues, they are putting
forth gayer petals, choicer blossoms, richer fruits, and so
they win in the race of life.   I must refer the reader to Mr.
Darwin himself, in order that he may learn what devices
the Orchid and other wonderful flowers have hit upon for
gaining a higher position in life; and no one who reads
intelligently his accounts of the way in which the blossom
has adapted itself to the insect, whose visits it covets, will
give up his study without feelings of profound admiration, and maybe of intense emulation, as he feels that
what the flowers can accomplish he also should do.

UR study would be incomplete were we not to make a few remarks on a branch of this subject which is of great interest : I mean that which relates to

CIRCULARITY IN LIFE.

Learned expositors and preachers have discoursed eloquently and pleasingly on the words of Solomon (Eccles. iii. 15), " That which hath been is now ; and that which is to be hath already been ; and God requireth that which is past." They have dwelt on the " law of circularity," and shewn how there is a constant revolving and circulation throughout the universe. Day gives place to night, and night again to day ; spring succeeds winter, and winter follows summer; the sun revolves, the earth is moving, the blood circulates, the smoke forms the cloud, the cloud brings rain, the rain produces nourishment and causes the trees to grow and enables us to make our fires and create more smoke, and so on and on we go, ever revolving, never standing still. Now, this law of circularity is beautifully seen in connection with the life history of the flowers and

D

plants. We begin at the beginning, and we find that the very simplest of all forms of life is the sphere, circle, or round cell. We find this form in the fresh water algæ, which abound in our pools, ditches, and water-butts. It is simply composed of a cell or sac filled with protoplasm. When it is about to evolve a new plant it divides itself into two parts, either by putting forth a little bud at its circumference which, gradually increasing in size, assumes the round form of the parent cell, and ultimately separates from it to become the ancestor of another similar form; or else by sending a partition across its protoplasm, which, in turn, is subdivided and four cells are formed from one. One of the most beautiful and interesting forms of microscopic plants is that known as *Volvox globator.* Men of science were a long time in discovering whether this minute form of life should be regarded as an animal or a plant. Curiously enough the law of circularity is here illustrated in two ways. Not only is the plant globular, as its name implies, but it is also, in common with many minute animals and plants, provided with little tails, whips, filaments, or cilia, by means of which it keeps itself in motion. Now the Volvox does not merely plough ahead, or move forwards and backwards, it is capable of revolving like a globe on its axis, and so the motion produced is of the most graceful description. The appearance of the

plant may almost be likened to a miniature cactus, perfectly round, on whose spines a spider has woven with
regularity and geometrical nicety its silken web. It will
often be seen that within the sphere itself little Volvoces
are revolving—wheels within wheels. Then there is the
Pandorina, which consists of a round cell in which a
great number of smaller cells or spores are to be seen.
These again are provided with cilia, which protrude
through the wall of the larger cell, and by their means
the plant is kept in motion. If you study closely the
minute forms of life which you place under the microscope, you will soon see that though their appearance
varies they are all composed of cells. Sometimes these
are simple spheres, at other times there are spheres within
spheres, but in yet other cases you cannot detect a sign
of rotundity. How is this? Because the cells have
contracted the habit of growing side by side, and
linking themselves together. If there is a single string
of cells the sides in contact with each other will be
flattened by pressure, while those sides which are
free will probably be rounded off. If, however, the
cells are in a mass, their angles will be of all shapes.
Yet it is easy to see that the original form is the round.
Hence, in plants of larger growth, while the cells have in
many instances retained their circularity, in others they
have had to give up their original shape under the in-

fluences which have been at work upon them. From cells of simple form we turn to those which are more complex, and here we notice that some are spiral and others annular. These cells are of vast importance to the life and development of a plant, and when you take up the study of plant physiology this will become more and more apparent. Then from cells we pass on to spores and pollen. When speaking of the smut (*Uredo*) found in wheat, we shewed how each spore was a simple globule or spherical cell, and if you take the spores of other fungi or those of ferns the same law will hold good. Exceptions prove the rule. Even the spore-cases, such as you see in the form of golden spots on the back of the common polypods, grow in circular patches, each patch being composed of a number of bags filled with miniature spores. These bags are composed of elastic cells which burst as the spores increase in size, and as we further examine the various parts of the plant we are constantly coming across further indications that the law of circularity is at work. This is seen, not less in the method adopted by the fern for production of young plants than in the constitution of the component parts of the plant itself. The micro-fungi, especially the cluster-cups (*Æcidia*) found on the Colts-foot, Barberry, Nettle, and Dock, are further delightful forms which may be studied. Not only do they usually form round patches or clusters, but each cup is round, and abounds in spores of a similar shape. Then the pollen of plants is further proof of the prevalence of this law. Here we have the pollen of the Hollyhock, perfectly round, and since it is designed for removal to another plant by insect agency, we find it studded with little hair-like appendages or projections which assist it in its endeavours to adhere to the object with which it is brought into contact. The essential organs of plants are also round. The pistil, ovary, and ovule each partakes of this character, which is no less conspicuous in the seeds themselves. Notwithstanding the fact that many fruits are anything but round, we are certain that the circular is the normal shape. We find it in our wild fruits, Hawthorn, Sloe, Crab, Yew, Ivy, Cranberry, Holly, Mistletoe, Briony, Hip, Acorn, Chestnut, and many others. So with the Gooseberry, Currant, Strawberry, Raspberry and Blackberry. The legumes give us round seeds as in the case of Peas, Vetches, Broom, Gorse, and their congeners. Larger fruits take the same shape, whether grown in bunches like the Grape, or singly as Oranges, Lemons, Fig, Longan, and the thousand and one edible and inedible fruits of this and other lands. Look too at the form taken by the flowers. While we have such highly developed modifications as the Orchids and papilionaceous plants, we find that all simple plants have round blossoms. The tropics abound in monstrous forms; here where life is quiet and the circular form answers most of the purposes of life we find it most common. Thus our Daisy and Primrose, Speedwell and Pimpernel, Ragwort and Campion, all the Buttercups and Rosaceous flowers, the Poppy family, the Crucifers and many others are round. Those which have novel shapes nevertheless shew in most instances that their normal form was the same as we see in simpler flowers, and you will find the calyx of a spurred Violet as regular in form as that of a Primrose. Once more, turn to the shapes assumed by the stems of plants, their roots and bulbs, and what do they teach us? The Crocus, Tulip, Snowdrop, Onion, Carrot, Parsnip and other flowers and plants have round bulbs or tap-roots, and even the larger trees send deep down into the earth a second stem or root similar in shape and form to that above ground. The pointed or tapering root is penetrating, and consequently capable of securing for the plant immense grip. Try to pull up a carrot or parsnip, and you will find that the tenacity of its hold is in direct proportion to the depth and size of its tapering root. Then you have the round stem of the Wheat, Oat, Barley and Rye, ; the Reed, Cane, Grass and Bamboo ; the Palm with its endless varieties ; the Fir and the Oak ; in fact, for the stem of plants and flowers the angular or flat is abnormal and unusual, the circular is common and normal. Think of the immense advantage which this form gives the plant over the square or many

sided shape.   In cases where (as in the lianas and other
creepers) the stem is flattened or angular a special end
has to be kept in view.   These plants do not stand alone,
exposed to the storm and blast.   They coil round the
stronger tree and live only as they find support from
others.   But the round stalk of the wheat and reed, and
the circular stem of the pine and palm, is peculiarly
adapted for exposure to changes of climate, atmosphere
and temperature.   It presents as little surface as possible
to the storm, the head is kept erect, and the interlocking
of the cells produces internal strength and support, while
there is greater elasticity in the motions of the plant when
swayed with the wind than could otherwise be acquired.
Thus the law of circularity is seen to be of exceeding
value in a variety of ways, and the thoughts which are
suggested may be followed out by each of us in the way most
congenial to our own mental and spiritual capabilities.

THE TRANSMISSION OF LIFE.

"NONE of us liveth to himself." This truth is abundantly illustrated by the way in which the plants vie with each other in their attempts to communicate their life to others. The methods adopted are very various, the end is the same. It would seem that the one object which the plants keep before them is the propagation of their race in the healthiest and strongest manner possible ; not merely the transmission of life, but such transmission in the way best adapted for the vigorous continuance and development of life. What a grand end to keep in view ! In this the ancient Greeks and the nobler plants are in unison. The end being ascertained, the question is raised—How shall it be attained? The object being set forth, the plan has now to be adopted. From the earliest times all plants were possessed of the power of passing on to their progeny the germs of life. The ability to do this was in many instances latent in each separate plant, as we see it still in the very lowest orders. They combined in one the male and female principles. But such a method of sustaining and transmitting life was subject to many disadvantages. Hence we find some plants possessed only of stamens while others had only pistils, and the agency of the wind was necessary to convey the pollen from the staminate flowers to the stigmas of the pistillate flowers. You see the Dog Mercury (*Mercurialis perennis*), for example, growing in dark green tufts under yonder hedge. That plant will illustrate our point. The Hazel and Oak have both kinds of flowers on separate twigs and branches, and as the former often grows as underwood, it is no unusual thing for the stamens of one plant to have their pollen wafted by the wind to the pistils of another plant, and thus cross-fertilization would ensue. But to supply enough pollen for such an undertaking involves great labour. If you gather the tassels hanging on the Hazel or Willow and shake them over a sheet of paper, you will be surprised to find what a quantity of yellow flour or powder falls from them. This is the pollen by means of which the pistils are fertilized, and fruit secured for the transmission of life. Gradually, so far as we can learn from the records of the earth, new forms of insect life sprang into existence, and side by side with their evolution we find new efforts made by the plants to utilize their agency and adapt themselves to the new and more perfect state and times in which they lived. We shall always be to some extent under correction in speaking about the flowers and insects of early times, for their preservation in anything like a perfect fossil state is next to an impossibility in the case of those with delicate organs, tissues, and colours. But we have the testimony of our own times, and the records of over two thousand years, since the study of plants was first commenced, and from the evidence derived from these two sources we can formulate a series of laws or arguments, which, applied to the study of fossil botany and entomology, will serve as a safe guide to much of the history of the past. We thus learn that in order to accomplish the great end of life the flowers sought the caresses of the insects and laid themselves out for decoying them. If honey glands were formed, honey-guides studded the petals, gaudy tints were dexterously applied to the corolla, the stamens and pistils assumed new positions and shapes, the pollen even, minute and dust-like as it is, adapted itself to the changed circumstances, and everything was done to ensure success. And how admirably have the various plans succeeded ! Such is the anxiety felt throughout nature that everywhere we find a planning and scheming going on, and

nowhere do we observe the process under more honourable conditions or favourable circumstances than in the
flowers. The youthful damsel does not trick herself out
with greater taste when she wishes to obtain the favour
and receive the addresses of a gallant beau, than does the
flower which seeks to win the loving favours of the butterfly or bee. But in the struggle for life all have not been
able to secure the attentions of these fitful visitants, and
they had run the risk of being left to die. Hence other
methods have been adopted by means of which to ensure
the transmission of life and the propagation and continuance of the species. The strawberry, for example, while
it possesses flowers which are fully equipped with all the
requisites for proper fertilization, has also received the
power of multiplying itself by means of runners. This
property is not confined to any one group of plants.
The Chinese have a flower (*Saxifraga sarmentosa*) which
has been introduced into this country under the name of
Aaron's Beard, whose multiplication may be effected in
exactly the same way. It is one of the many sisters of
our London Pride (*Saxifraga umbrosa*) which is propagated by stoles or offsets. You will often see it in cottage
windows throwing out its trailing runners, on the end of
which young plants appear. This peculiarity has led to
its being called Spider Plant and Mother of Thousands.
So we have the Crowfoot or Creeping Buttercup (*Ranun-*

culus repens), a most troublesome plant, and one which every gardener and farmer detests. This too has the same power, and right vigorously does it use it. The same may be said of many other plants, such as the Silver Weed (*Potentilla anserina*) and especially the Creeping Cinquefoil (*Potentilla reptans*). The Bramble has the power of propagation by means of roots thrown out at the end or middle of a branch which has been reclining on the ground, and by following out this idea gardeners are able to make "layers" of Gooseberry, Currant, and other bushes, shrubs and trees. Laurel plantations are thickened and rendered uniform by pegging their young branches into the ground in such a way as to enable them to take root and so form fine vigorous plants. Then we have suckers springing from the Raspberry, Asparagus, Mint, Bamboo, Sumach, and other plants. These are the result of the roots bearing buds which eventually send up a stem exactly like, but at a distance from, the parent. The American Blackberry is now being cultivated in this country, and year by year the old stems are cut down, as we do those of the Raspberry, depending for new fruit on each year's growth. Many plants are multiplied by means of tubers, as is the case with our old friend the Potato, and the "pomme de prairie," or esculent Psoralea, a farinaceous root found in Canada, which affords during winter a very acceptable and nutritious food for the natives of the regions where it grows. Some, again, like the Daisy, London Pride and Primrose, may be multiplied by dividing the parent plant, so that from one clump of Saxifrage enough plants may be obtained for planting an entire border. Then there is the bulb and corm. Many of our choicest flowers depend for the transmission of life upon the young bulbs they produce, and we seldom think of looking for the seed of the Snowdrop, Crocus, Tulip, or Hyacinth, since they have the more substantial root formation to fall back upon. Many plants possess the property of starting afresh on the journey of life by means of roots thrown out from a tiny portion of stem or leaf. Cuttings may therefore be taken from Gooseberries and Currants, Pinks and Carnations, Fuchsias and Geraniums, and many other plants, which, with due care and attention, will soon strike and grow into useful and vigorous plants. Even the portion of a leaf of some plants possesses this power and vitality, and so we find an infinite variety of methods adopted by plants for the successful attainment of the great end of life. I have spoken of the fertilization of plants. Here it may be well to remind the reader that the greater efforts we see put forth by plants to secure the fertilization of their flowers, the more perfect will the seeds become. The microscopic study of the smallest seeds of insect fertilized plants is full of interest. It is the seed which in the normal plant secures the transmission of life. It is interesting to observe that as soon as a plant has succeeded in securing its object it dies away, leaving its fruit or seed to carry on its work. Of the great variety of seeds and fruits, it is impossible here to speak ; but the wonderful provision made is fraught with lessons. As a rule, that part of the fruit or seed which carries the germ of life is produced in a case or covering. Thus the pea and bean, together with other legumes, grow in pods, while the seed itself is carefully covered over with a coat of what we might call vegetable vellum, which falls off when the germ begins to grow and the cotyledons swell. In the case of edible fruits, the seed is usually in the centre, and this in many cases is covered over with a hard shell, which divides when the seed begins to grow, and ceases to need its protection and help. Families of plants usually adopt like methods throughout the community ; but this rule is not absolute. While, for example, the Orange family bears seeds or pips in the centre of the fruit, the Rose tribe has more than one method of placing its seeds. In the Crab we have one method, in the Hip and Haw another, and in the Strawberry another. Some provide for the agency of birds, hence the pulp of the Cherry, Hips and Haws, Sloe, and Holly ; some rely upon the wind for dissemination, whence we find the "keys" of the Ash and Maple with their delicate wings, and the winged seeds of many other plants. The pappus on the Thistle and Dandelion seeds also serves the same purpose, and on being caught by the passing breeze the whole is wafted to some distant spot and deposited, seed downward, ready for germination. Very ingenious devices are found in operation among some plants. The Balsam goes off with a bang when its seeds are ripe, as everyone has observed, and by this means the seeds are sent far and near. The wild Geranium has a similar method of scattering its fruit, and the appearance of the seed vessels after they have discharged their contents is most remarkable. They can be seen by any hedgerow, and should be studied by all who wish to know how Nature provides for herself. We may say that these plants possess within their capsules a series of elastic catapults, with which they throw their stones in every direction. The Burdock, Cleavers (*Galium Aparine*), and other plants, are provided with curved hooks, by means of which they fasten themselves upon the wool of sheep or other passing objects, and so travel from place to place. Thus in one way and another the great end is gained, and the flowers hand on to others the life which they have received. One thing strikes us as worthy of notice. While here and there a plant, like the Orchis, is content with simply providing for one or two descendants, some aim at great results. True, they do not always succeed. Thousands of Acorns become the food of squirrels and other animals to every one that becomes a tree; but in many instances the attempt to scatter life broadcast succeeds, and that plant which can send forth the greatest number of well-developed and well-protected seeds, bulbs, runners, or suckers, stands the best chance of winning the race. Is not the thought suggestive with reference to the work of the Christian and the Church? Every one of us may learn that while it is a great thing to be able to do a little in the way of transmitting the life which we have received to others, the more earnestly we work, the more liberally we sow our seed, the greater chances have we of succeeding in our endeavours. " In the morning sow thy seed, and in the evening withhold not thy hand; for thou knowest not whether shall prosper, this or that, or whether they both shall be alike good" (Eccles. ii. 6). Thus may we be ever found energetically seeking to transmit life to others.

THE AUTUMN OF LIFE.

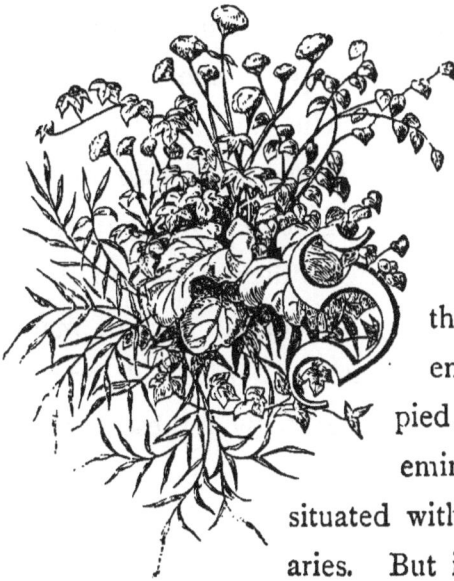

HERWOOD Forest is famous on account of its association with Robin Hood and his followers in former times, as well as from the fact that many anci-ent mansions, still occu-pied by dukes and other eminent personages, are situated within or near its bound-aries. But it also has a name for its beautiful trees, including the Major, Parliament, and Greendale Oaks. I was driving through the forest recently, and was struck with two things. First, one could not fail to admire the glory of the autumn tints; but, on the other hand, the "autumn of life" was sadly apparent in the decayed and weather-beaten forms of many of the noblest trees. Here, as we drive along the private roads around Edwinstowe, we see oak after

E

oak with its topmost branches gone, its head carried away, its trunk decayed, or its bole twisted. Many a heavy storm in the good old times, when winters were probably more severe than we know them, has swept across this beautiful forest; and it is not to be wondered at that trees which have seen four, five, and perhaps more, centuries of summer and winter changes—wind and storm, sunshine and shower—should begin to give proofs of being beaten. The sight is depressing. One almost wonders at times why the owners do not cut them down as cumberers of the ground. But that would be Vandalism indeed. So long as they can stand at all they have a purpose in life. Around us we see many relics of the past, and the tottering, enfeebled form of the aged patriarch is very touchingly represented by these old trees. The shaking head and tremulous hand tell their own tale. They have stood many a storm, and grappled with many a difficulty; but in the long run, time and trouble get the victory, and the strongest of us has to give way. But with reference to the autumn tints, how suggestive are they of thought respecting life! The autumn of 1884 was one of the grandest that has been known for years. I had the opportunity of seeing the country from north to south and east to west, during the months of September and October, and certainly I never saw anything more beautiful in my life than the

changing tints of the foliage, and the bright colours of
the leaves. On leaving Stonehenge for Salisbury, through
the villages we passed scenes of surpassing grandeur.
The view from Carisbrooke Castle, Ventnor, Boniface
Downs overlooking Bonchurch and Apuldercombe, Shank-
lin Chine, and other places in the Isle of Wight; the
New Forest, seen from an eminence above Rufus' Stone ;
the avenues of trees in Oxford, the forest foliage in
Nottinghamshire and Yorkshire—these alike told the
same story. And what was it ? The lesson which, I
think, we might learn is this—that the autumn of life
depends upon the kind of spring and summer we have
had, and the way in which we have improved it. The
summer of 1884 was of such a kind that the trees became
stored with sap, and developed strong, healthy leaves.
The autumn came gradually, not with intense cold for a
night or two, followed by wintry blasts of a wild and
destructive nature. Slight frosts took off the " greenth "
of the leaves with a delicate touch; warm, sunny autumn
days threw light and shadow on the ever-changing scene,
and the leaves clung to their posts with brave and cheerful tenacity. They had enjoyed a most favourable summer
season, and the later days of their existence were not
merely peaceful, but even gay and joyous. Such be the
autumn of my reader's life !

How many a time has reference been made to the

falling leaves as an illustration of human life, or perhaps
I should rather say of human decay. " We all do fade
as a leaf," says the prophet (Isaiah lxiv. 6). Sometimes,
indeed, the leaf falling is an indication of calamity. The
Lord on one occasion (Jeremiah viii. 13), threatens the
Vine and Fig tree with barrenness, and adds, "the leaf
shall fade," a token that health and vitality has gone.
But such a calamity it is not our intention to dwell upon
here, it is sufficient that we refer to the matter and suggest the subject as one which we may each meditate upon
with profit. But autumn is not a calamity. It must
come in the necessary order of things. It is as needful
as is the summer, or the spring. It is also the season of
fruit bearing. In the spring-tide the seed is sown, in
summer the plant gathers strength, produces blossoms,
and fertilizes its flowers, but in autumn they come to perfection and are gathered. While the harvest fields are
filled with golden grain, the orchards are as beautiful as
they were in May, when rosy blooms were on every
bough. The fruit has changed its colour, and Apples,
Pears, and Plums, are ready to be gathered and stored
away for future use. What could we do without autumn?

To return to the question of tints. I remember reading the following sentences in the works of a well-known
American Divine, whose brother I had the honour of
knowing as one of the most devoted but unobtrusive

workers in the Chinese Mission field. "For several autumns," says Dr. Talmage, "I made a lecturing expedition to the far West, and one autumn about this time saw that which I shall never forget. I have seen the autumnal sketches of Cropsey's, and other skilful pencils, but that week I saw a pageant two thousand miles long. A grander spectacle was never kindled before mortal eyes. Along by the rivers, and up and down the sides of the great hills, and by the banks of the lakes, there was an indescribable mingling of gold, and orange, and crimson, and saffron, now sobering into drab and maroon, now flaming up into solferino and scarlet. Here and there the trees looked as if just their tips had blossomed into fire. In the morning light the forests seemed as if they had been transfigured, and in the evening hour they looked as if the sunset had burst, and dropped upon the leaves. In more sequestered spots, where the frosts had been hindered in their work, we saw the first kindlings of the flames of colour in a lowly sprig; then they rushed up from branch to branch, until the glory of the Lord submerged the forest. Here you would find a tree just making up its mind to change, and there one looked as if, wounded at every pore, it stood bathed in carnage. Along the banks of Lake Huron there were hills over which there seemed pouring cataracts of fire, tossed up, and down, and everywhere by the rocks.

Through some of the ravines we saw occasionally a foaming stream, as though it were rushing to put out the conflagration. If at one end of the woods a commanding tree would set up its crimson banner, the whole forest prepared to follow. If God's urn of colours were not infinite, one swamp that I saw along the Maumee would have exhausted it forever. It seemed as if the sea of Divine Glory had dashed its surf to the tip-top of the Alleganies, and then it had come dripping down to lowest leaf and deepest cavern." This is a word picture which is worthy of a golden setting. By its side should be placed another, painted by that interesting and facile writer on flowery subjects—Shirley Hibberd. In his delightful book, entitled "Brambles and Bay Leaves," you will find a chapter on *A Season of Brown Leaves*, which is indeed a popular and pleasing sermon on our text—the autumn of life. The various thoughts suggested by this subject are there worked out just as we should like to see them done, and after reading Mr. Hibberd's words one feels afraid to venture on the same track. Let us cull one paragraph from his book in place of an attempt to say the same thing it our own words. "The history of man, no less than the history of nature, teaches this lesson of evolution. Wrapped up in the oval bud of spring are the blossoms and fruits of the summer ; and in the impulsive heart, beating in harmony with the instinctive nature of the primeval man, are enfolded the acts of his illimitable successors. The shepherd-life, with its simplicity and peace, is seen again in the radiant face of the infant, and the violet tenderness of the spring. The age of chivalry, with its costly pomp, its clang and clash of arms, its great deeds of daring and sacrifice, break out in the hours of individual passion when manhood has not yet set its seal on the brow, and when the outward semblance of heroism is mistaken for the supporting and sustaining ardour which springs from manly determinations. The first flush of summer has it, too, when the fruits are yet unripe, and storms dash in and out between the leaf-laden branches. But the autumn and the browning leaf must come, and it is already here around us. Who, then, is worthy to die—worthy as the leaves are, all of whose duties have been fulfilled? Who is worthy to convert body and soul into a soil for the growth of the next generation of men, whose bodies are to be formed out of the elements of ours, whose spirits are to be fed with the aims, and hopes, and knowledge we have nurtured, and which we must bequeath to them by an inevitable necessity? Who among us has been living all these years in vain, watching the greening and the browning of the leaves, without taking heed that his autumn must come, and that winter must heap snows on his tomb, as upon the graves of fallen leaflets?"

The study of nature's operations in autumn is not less interesting than that which relates to the season of the opening bud and flower. It is perfectly marvellous how carefully every step in life is taken, and what provision is made for each change of form and season. When the leaves have done their work, acting to the plant as both lungs and stomach to the animals, they are gradually detached from the twig which has borne them, by a most ingenious process. The separation is produced by means of a joint or articulation, which gradually forms between the stem on which the leaf is growing, and the stalk or petiole of the leaf itself at the point of contact with the stem. A layer of cells having been formed at the base of the leaf stalk, a clean scar is left on the stem when the leaf falls off. If you break off a leaf from an Ash tree in spring you will have ruthlessly to sever the various cells and tissues at the point of rupture. If in summer you perform the same act it is likely that the leaf will come off with a certain amount of ease, and the wound will be slight, the scar having already commenced to form ; but if in autumn you touch the leaf which is ready to fall, you will see that all that portion of the twig which was covered by the leaf stalk has grown over in a regular manner, so that no injury is done either to the leaf or the stem in its detachment. A beautiful picture of life ! Ruthless is the hand that snatches away the babe and youth. The middle-aged have already begun to undergo a loosening process, albeit all unconsciously to themselves; but when the autumn comes the falling leaves are watched with something of pleasure as we reflect that they have lived out their days and done their work. As with the leaf, so with human beings. The fall of the leaf is not accidental. It does not result simply from change of season or temperature, or as the result of wind, rain, or frost. It is a regular process, and death is but a part of life. This process even commences with the first formation of the budding leaf, and only when the organ has done its work is the act completed. Ruthless forces may produce a premature fall, but if temperature, and situation, and circumstances of every kind favoured the plant, its leaf would eventually run its course, a scar would be formed before the wound was exposed, and the now useless member would be silently dropped.

If the reader will bear with a few words relating to the science of the subject, we will try and ascertain the cause of the tints of autumn leaves. It is well known that they vary in different plants : yellow, brown, and red, are favourite colours. Thus in the Birch and Willow we usually find a yellowish tint. In the Vine the colour of the autumn leaf is red, and curiously enough the depth of leaf-colour is in proportion to that of the fruit. Thus

black Grapes grow on Vines whose leaves in autumn are
of a deep colour; the leaves of the red Grapes are
lighter, while those of the white varieties have leaves of
reddish hue or even yellow. Who has not observed the
gorgeous colours of the Virginian Creeper? A row of
Beech trees, such as we often find in various parts of
Buckinghamshire, is a sight not to be easily forgotten if
seen in their autumn dress as they appeared last year.
Now the leaves of plants, in common with other organs
possessed of a green colour, are pervaded with a peculiar
substance known as chlorophyll or leaf-green. This
substance generally exists in a granular form in the cells
of the plant. When oxydation takes place in the leaf the
colour of this substance changes, but some have supposed
that the tints are due to the presence of pigments or
colouring matter which is distinct from chlorophyll.
When water is absorbed by the roots of a plant, and
carried up into the leaves, a greater or less amount of
mineral matter finds its way in solution through the
various cells. In the course of time the moisture is ex-
haled but the substances absorbed are left behind, and
fill up the walls of the various vessels and cells of which
a plant is formed. Hence in autumn the leaves contain
a much larger amount of mineral matter than in spring.
This prevents the growth of the leaf, and leads to its
gradual decay.

Dr. Talmage remarks that most people find in the words " We all do fade as a leaf," a vein of sadness, but he truly adds, that there is a string of joy to the harp. Like the leaves, we fade gradually. The trees are not denuded in a day, but first a leaf here, and then another there falls, till at length all have disappeared. The ties and cords are gradually loosened till the time is ripe for dissolution. Like the leaves, we die both to make room for others and to assist their growth by the stimulus of our own vigorous life. What a grand heritage do we enjoy on whom the end of the ages have fallen ! What stores of rich material into which to strike our roots ! What libraries, laboratories, museums, and galleries ! Our position is like that of the Sequoias which have grown up in the soil of a primeval forest, and we too must leave behind us our quota of material for the men of the future to live upon. The leaves fade and fall only to rise again. "All this golden shower of the woods is making the ground richer, and in the juice, and sap, and life, of the tree, the leaves will come up again." So the law of sacrifice is ever in force, and in the autumn of life we are but obeying this law, and yielding up ourselves for the good of the race.

THE TREE OF LIFE.

ANY curious facts have, during the past few years, been brought to light through the study of folk-lore and comparative mythology. One of the most interesting results is to be found in the community of thought, feeling, and action of men in every part of the globe and in all ages. Men placed in like circumstances, though far apart, think and act in a similar way, from which we argue a common nature. But many discoveries have been made which, if they do not actually prove by themselves the original unity of the human race, the possession of a beautiful homestead or grove by our early ancestors, the existence of trees and fruits possessed of special qualities, the fall of man through sin, and many other matters, at any rate they throw interesting rays of light upon these topics. Much has been written about the trees of Paradise, and especially the Tree of Life. The subject is as old as the Bible itself, and its elucidation dates from the times of the earliest commentators. Not only did sacred writers of later times, such as Solomon, David, and St. John, speak of a Tree of Life, but the Apostolic Fathers and early Divines loved to dwell with vivid and fertile imagination upon the same theme. It has, however, fallen to the lot of writers during the present century chiefly to bring together the notices of the Tree of Life and the common beliefs of the most widely scattered people respecting it. Until recent times it was scarcely known that Arabs, Hindus, Malays, and Polynesians, together with many other peoples, regarded certain trees as possessed of peculiar properties such as those attributed to the products of Eden ; but we have now ascertained that the idea is well nigh, if not quite, an universal one. It is useless to speculate respecting the nature of the trees found in Paradise ; but it may be interesting to notice what trees have received special honour in distant lands, and then see what lessons are taught by Scripture. In some lands, as would readily be imagined by those who know the fruit, the Banana or Plantain has the honour of being associated with Paradise. In Johnson's edition of old Gerarde's Herbal (A.D. 1633, generally quoted as *Ger. Emac.*), we find three illustrations of this fruit and plant : viz., first, Adam's Apple tree (*Musa serapionis*) ; secondly, Adam's Apple (*Musæ fructus*) ; and thirdly "an exacter figure of the Plantaine fruit." The author states that in his day " some have judged it the forbidden fruit; others the grapes brought to Moses out of the Holy Land." He adds, " The Grecians and Christians which inhabit Syria, and the Jewes also, suppose it to be that tree of whose fruit Adam did taste, which others think to be a ridiculous fable." On the strength of this idea, we find that one species of Musa has been nominally associated with Paradise. Thus we read : " The two most valuable and best known species of Musa are the Adam's Apple or Plantain (*M. paradisiaca*), and the Banana-Plantain (*M. sapientum*) ; the latter is a denizen of the New, the former of the Old World. Indeed, as the specific name *paradisiaca* imports, a notion was entertained by the old botanists that this was the forbidden fruit of Eden. . . . The native Indians use the leaves as plates, dishes, and napkins ; and those persons who believe the fruit to be the forbidden Apple of Paradise have also adventured the groundless surmise that the large leaves of the Plantain were the so-called fig-leaves of which our first parents made their aprons." One thing is certain, that if Banana leaves, such as we see to-day in the East, had been employed, they would not have needed sewing together ; for if a hole were cut through one end and the head thrust through, the leaf would easily reach to the knees or even feet of ordinary mortals. But while the usefulness of this tree makes it unlikely that it would be wholly tabooed or forbidden, we find Canon Kingsley suggesting that it may for ages have been regarded as a sacred tree. He says ("At Last," II. 271), "It is wild nowhere now on earth. It stands alone and unique in the vegetable kingdom, having distant cousins, but no brother kinds. It has been cultivated so long that, though it flowers and fruits, it seldom or never seeds. The only spots in which it seeds regularly are the Andaman Islands in the Bay of Bengal." Out of the hundreds that I have eaten, always of course avoiding the cross which may be seen on cutting it through, I have never come across its seeds. But the Palm has received even a larger share of notice than the Banana. Tradition asserts that the fruit of the Date-Palm was one of the three things which fallen Adam was allowed to take with him when driven from the Garden of Eden. These dates were eventually planted, and from them all the Palms in the world have sprung. The sacred tree which we find so frequently in the Assyrian sculptures appears to be a traditional form of the Date-Palm, a tree which in Egypt and Arabia particularly has long been regarded with profound esteem. As Mr. King truly says ("Sketches and Studies," p. 37), "It has a special beauty of its own when the clusters of dates are hanging in golden ripeness under its coronal of dark-green leaves. It is figured as a Tree of Life on an Egyptian sepulchral tablet, certainly older than the fifteenth century, B.C., and now preserved in the Museum at Berlin. Two arms issue from the top of the tree, one of which presents a tray of dates to the deceased, who stands in front, whilst the other gives him water—'the water of life.' The arms are those of the goddess Nepte, who appears at full length in other and later representations." The male flowers of the Date-Palm grow on one tree, while the female grow on another. Owing to this (diœcious) character, the fruit often fails or degenerates in bad seasons, on which account the Arabs, when necessary, cut the pollen bearing spikes of bloom and hang them over those bearing pistils to ensure their fertilization, and thus a choice supply of fruit is secured. A festival is held at this season of the year which is known as the Marriage of the Palm. It has been asserted that the tribes of people are so familiar with the importance of the due impregnation of the pistils with the pollen of the staminiferous trees, that one writer states that the threat to destroy all the male trees growing in a certain region led to an invasion being warded off. " I remember (says Kœmpfer) it happened in my time that the Grand Signior meditated an invasion of the city and territory of Bassora, which the prince of the country prevented by giving out that he would destroy all the male Palm trees on the first approach of the enemy, and by that means cut off from them all supplies of food during the siege." The reader may be referred to the learned and valuable *Mythologie des Plantes* of Comte Angelo de Gubernatis for the further elucidation of the subject connected with the Banana, Palm, and other trees. We may now remark that the next tree which has been associated with the garden of Eden is the Fig. This has partially resulted from the statement that the culprits, on discovering that they had sinned, sewed Fig leaves together in order to make gar¬ ments. Thus in an Egyptian scene which Rosellini has figured from the sepulchral tablets, we find several generations of a distinguished family receiving nourishment from the Tree of Life, one of the Fig trees (*Ficus syca-morus*) being the type selected. The goddess Nepte rises from the top of the tree with a tray of Figs in one hand, while she pours a stream of water from a vase held in the other. The various peoples of the East adopt dif-ferent types of Fig, that held sacred among the Hindus being the Peepul or Bo-tree (*Ficus religiosa*). Both this and the Banyan are regarded in India as the Tree of Knowledge, and with them Buddha and other renowned personages are intimately associated. The Rabbins, fond of multiplying legends and producing marvellous stories, "describe the Tree of Life as being of enormous bulk, towering far above all others, and so vast in its girth that no man, even if he lived so long, could travel round it in

F

less than five hundred years.   From beneath the colossal base of this stupendous tree gushed all the waters of the earth, by whose instrumentality nature was everywhere refreshed and invigorated.   Regarding these Rabbinic traditions as purely mythical, certain commentators have looked upon the Tree of Life as typical only of that life and the continuance of it which our first parents derived from God.   Others think that it was called the Tree of Life because it was a memorial, pledge, and seal of the eternal life, which, had man continued in obedience, would have been his reward in the Paradise above. Others, again, believe that the fruit of it had a certain vital influence to cherish and maintain man in immortal health and vigour till he should have been translated from the earthly to the heavenly Paradise."   ("Plant Lore," by Richard Folkard, p. 13.)   It has been argued with considerable ability that the Banyan was the tree in the midst of which Adam and Eve endeavoured to conceal themselves.   This famous tree produces aerial roots which strike downwards from the branches, and, reaching the ground, form new trunks and infant trees which might be separated from the parent and yet retain their vitality. Such a tree, as all who have lived in the East are aware, forms, in the words of Milton—

> "A pillared shade
> High over-arched, with echoing walks between."

The Tree of Life is so intimately associated with the
Tree of Knowledge, that in noticing one we are obliged
to refer to the other. We are in the habit of speaking of
the Apple as the forbidden fruit, and the " Adam's Apple"
found in the throat has received this designation from the
legend, that when the forbidden fruit was partaken of by
Adam a piece of it stuck fast, and produced the protub-
erance. In China, however, it is an Olive which has
become fixed in the throat. The Orange, Citron, Pome-
granate, Grape, and other fruits have been described as
having each a strong identification with the forbidden
fruit. Speaking of the Citron, Gerarde says that among
the vulgar sort of Italians of his day, the fruit was called
Pomum Adami, or Adam's Apple; "and that came by
the opinion of the common rude people, who thinke it to
be the same Apple which Adam did eate of in Paradise,
when he transgressed God's commandment; whereupon
also the prints of the biting appear therein, as they say :
but others say this is not the Apple, but that which the
Arabians do call Musa or Mosa," *i.e.* Banana. Among
the Indians of Orinoko the Moriche Palm is held sacred,
whence it was named by the Romish Missionaries the Tree
of Life. Kingsley tells us that, according to the Tama-
nacs, after a great deluge which swept man from off the
earth, a man and woman, the sole survivors of the human
race, cast the fruit of this Palm behind them, and watched

till from the seeds a host of men and women rose up to re-people the world. Many such stories are to be found in the mythologies of Greeks, Romans, Scandinavians and others. Speaking of Wisdom, Solomon says (Proverbs iii. 18), "She is a tree of life to them that lay hold upon her; and happy is everyone that retaineth her." But St. John in the Apocalypse has two most interesting references to the subject of this tree. He is instructed to say to the Church of Ephesus (Revelation ii. 7), "To him that overcometh will I give to eat of the Tree of Life which is in the midst of the Paradise of God." And he further adds (ch. xxii. 2) that, "On either side the river was there the Tree of Life, which bore twelve manner of fruits, and yielded her fruit every month; and the leaves of the tree were for the healing of the nations." On the former of these passages, Archbishop Trench has an able note which is so much to the point that I give its substance to the reader. In "the Tree of Life," there is manifest allusion, says he, to "the Tree of Life in the midst of the garden," mentioned in Genesis ii. 9. The tree which disappeared with the disappearance of the earthly Paradise, reappears with the reappearance of the heavenly, Christ's kingdom being in the highest sense the restitution of all things. Whatever had been lost through Adam's sin is won back, and that too

in a higher shape, through Christ's obedience. Hence
the poet says :—

> " In Him the tribes of Adam boast
> More blessings than their father lost. "

That the memory of this tree had not in the meantime
perished, we gather from a number of references in the
Book of Proverbs. In addition to that already quoted
we find (xi. 30) "The fruit of the righteous is a tree of
life." So (xiii. 12) "Hope deferred maketh the heart
sick : but when the desire cometh it is a tree ot life.
Lastly, "A wholesome tongue is a tree of life" (xv. 4).
The Rabbins, of course, knew a great deal about this
tree of life, as we have already seen. Its boughs, they
said, overshadowed the whole of Paradise. It had five
hundred thousand fragrant smells, and its fruit as many
pleasant tastes, not one of them resembling the other.
" To eat of the Tree of Life " is a figurative phrase to ex-
press participation in life eternal. " Blessed are they that
do His commandments, that they may have right to the
Tree of Life" (Rev. xxii. 14). More than once in the
Apocrypha do we read of this tree. Thus (2 Esdras ii.
12) "They shall have the Tree of Life for an ointment of
sweet savour." Esdras asks what profit it is to a man
"that there should be showed a paradise, whose fruit
endureth for ever, wherein is security and medicine, since
we shall not enter into it?" In Ecclesiasticus xix. 19,

we are told that the knowledge of the commandment of the Lord is the doctrine of life; and "they
that do things that please Him shall receive the
fruit of the tree of immortality." We meet with echoes
and reminiscences of this Tree of Life in the mythologies of many nations ; or, if not actual reminiscences
of it, yet reachings out after it, as in the Yggdrasil
of our own northern mythology, and still more remarkably
in the Persian Hom (Haoma). This Hom is the king of
trees, is called in Zendavesta, the Death-destroyer. It
grows by the fountain of Ardinsur—in other words, by
the waters of life ; while its sap drunken confers immortality. This is the sacred Soma of the Hindûs, of whose
virtues we read in the sacred literature of the East.

Here we must close our study of the teaching of the
plants and flowers respecting life. The subject may be
further followed up by such as desire to do so in the
many able works which have been already written. Our
gleanings have been sufficient to shew what a wide field
for profitable meditation Nature presents, and he who
rightly enters upon the study will not only find it more
and more engrossing as he proceeds, but will also be
led to a truer and profounder devotion to the great Giver
of life, whose finger and handiwork we can everywhere
trace. We shall probably never get nearer the explanation of the mystery of life, for that appears to be a subject

altogether beyond the reach of science ; but we shall see that its possibilities are limitless, while from the infinitely small to the infinitely great it is equally perfect and beautifully developed. May the lessons with which so profound a subject is fraught be deeply written on our hearts, and lead us each from Nature up to Nature's God.

BOOK II.

Œ̓e Ministry of Flowers

RESPECTING THE EVILS OF LIFE,

" Degeneration follows, and with it all sorts of vegetable vices and dodges to gain a bare living, or for hanging on to life."

Dr. E. J. Taylor.

BOOK II.

The Evils of Life.

IMAGINE the reader saying to himself—"Pray, what can the flowers teach us about life's evils and vices?" Have you never been startled by the fact that the curse which has fallen upon man seems to rest on the whole creation? Have you never observed that, just as the animals — from man downwards to the lowest link in the chain of life—" delight to bark and bite," to snarl, cheat, deceive, and even devour? So the same evil qualities are to be found fully developed in the plant world. If this fact has never struck you, there is before you a wonderful mine for research, and a most extensive field for your careful observation. Let us now come to the consideration of some of the facts in the history of flowers and plants, which show us life in its darker moods. And let me, in the first place, speak of the general

CORRUPTION OF LIFE

that abounds. As we have in every grade of human society, spots of pollution and defilement; as every rank in life has its corrupting influences ; as even the sun has its spots—so the flowers have to contend with influences of an equally degrading and mischievous character. These vices are not merely such as stain the pure petals, and spoil the beauty of the plant : they even eat out its life, and destroy its vitality, leaving it a perfect wreck. Come with me to the field of waving corn, which we see in the distance yonder, and I will explain more fully what I mean. How bright and cheerful the appearance of the field, as the wind gently rustles in the corn-ears, and makes them bend in rhythmic waves, which undulate from one end of the plot to the other ! We pronounce the field—" white already to the harvest "—a grand success, and are sure the farmer will be well repaid for his outlay. But wait a bit. The farmer himself is coming toward us,

and ever and anon we notice that he stops to examine
the ears of corn, and looks disappointed and sad. We
shall cheer him if we congratulate him on the fine crop,
maybe, so we at once commence a conversation on the
subject. He listens to us for a while, and then observes,
"You cannot have given the field a very close examination. I regret to say that the crop is largely affected
by smut, rust, and blight. In fact, I have not seen a field
in such a state for years, and this is all the more grievous
since it is a long time since the corn ripened so well, or
bid so fair to return a good harvest." Now, we shall
judge of the farmer's feelings in proportion to our knowledge of the things he has been describing. It is quite
likely that he does not know how blight differs scientific-
ally from smut, or what is the gulf which separates brand
from blast ; but of this he is sure, that since these things
are in the wheat, his sample is spoiled. Now, it is one
of the triumphs of the microscope, that it has revealed to
us the history and mystery of these various plague-spots ;
and to their study we may now with profit, and I trust
with pleasure, turn. Let us take home a few of the straws
and ears which the farmer has gathered, in order to show
us what he means, and see what we can make out of these
red and black spots. We will first examine this ear of
corn, which is attacked with smut. Take a slip of glass,
and shake some of the black dust from the ear upon it ;

then place it under the microscope. If your lens is sufficiently powerful, it will reveal to you some startling facts. You see that the dust, which you have rubbed down to the finest dimensions—so that you can scarcely see it with the naked eye—is in reality made up of multitudes of tiny globules or spheres, each one of which is as perfect in its rotundity and finish as the brightest star or the choicest grape. Now, herein lies the mystery : that each of these tiny specks—thousands of which are huddled together in one grain of wheat—is possessed of life, and is capable of producing a new plant of its own order. You will observe that the globules are of the simplest character. They are not held together by a common receptacle—as peas, for example, are by their husk—but they just adhere together till separated by a gust of wind, or a passing shower. In the case of the kind under examination, the grains of wheat afford sufficient protection to these minute fungi, which formerly bore the botanical name of Uredo, from a Latin word *uro*, I burn, on account of the corn which is affected by some species appearing as if scorched and burnt up. This is the simplest kind of vegetable fungus, consisting as it does of a mere collection of spores, unconnected by a receptacle or ascus. They are interesting on this account, seeing that sporidia—which are really naked and unprotected by a native or homogeneous covering—occur but rarely among the fungi, whence the fungologist relies for guidance in

.

his classification upon the presence or absence of a spore-bearer. Two great primary divisions exist—the first, known as Sporifera, being represented by our friend, or rather enemy, smut; the second, called Sporidifera, including all those species whose spores are enclosed in cases or asci. It has been remarked ("Outlines of Botany," p. 183), that "blight, like brand and blast, is a term which has been popularly applied to all these small fungi indifferently, and is indicative of the former opinion—still entertained by many—that the plants affected by them have been star-struck, burned, or blasted by some atmospheric or planetary influence : names which were given in ignorance, thus being retained long after the error has been detected, and the truth revealed." It must be borne in mind that though the real character of these minute parasites has only recently been discovered, their actual existence in ages long since past can be easily proved. Thus "the fossil Lepidodendra, of the carboniferous period, is found riddled and perforated by the minute inter-penetrations of a parasitical fungus not distantly re-lated to our too well-known potato-disease germs " (Dr. Taylor). Further, in ancient times, "when the true nature of these visitations was unknown, religious ceremonies were chiefly resorted to for the purpose of averting the presumed anger of heaven, or of appeasing a supposed offended deity. The Romans, in consonance with their established customs, deified the cause of their dis-

tress ; and, after the apotheosis of the brand, it was wor-
shipped under the name of Rubigo. The Robigalia were
propitiatory sacrifices and feasts, instituted in honour of
the god ; they were held in the beginning of May, at
which time Rubigo was besought to let the corn escape
his fearful blasts." Thus, as early as the time of Ovid,
the farmer's enemy was known and feared. The question
may be asked—" How does the smut propagate itself?
If it is but a simple spore, or round speck of cellular
matter, how does it grow ?" Experiments and investigations have led to the conclusion that they are conveyed
through the cells and sap-vessels of the plant, along with
the moisture which the roots drink in, till they pervade
the most intimate structures of the plant attacked, and
eventually lodge in the parenchyma of the stem or ear,
where they produce an innumerable progeny. The attacks
of the smut are not confined to the grain. The insidious
intruder affects the culm, husk, and leaf as well, distorting
and disfiguring the whole plant, till it eventually becomes
dry and shrivelled, and looks as dirty as if it had been
employed in sweeping a chimney. As more will be said
on the subject of these micro-fungi by-and-bye, I will
here close the consideration of this painful subject,
merely suggesting that the facts which have been produced
convey a vivid idea of the corrupting nature of sin, and
teach us that, in its minutest and most insidious forms,
its attacks are to be dreaded and averted.

THE TRIUMPH OF EVIL.

HIS is another matter concerning which many illustrations may be found in our studies of plant-life. A native of Kent or Sussex, brought up in the midst of the famous hop-gardens, will often have admired the wonderful property possessed by the hop of winding itself around the poles set at intervals by the little "hills," as the beds are called in which the cultivated plants are trained. This peculiar property is not by any means the exclusive monopoly of the Humulus. The Scarlet Runner (bean) also possesses it, and such flowers as the Convolvulus raise themselves by its means to a very great height. The Honeysuckle is able even to clasp and climb the stem of a large tree. But the point upon which I wish to dwell is this : In other lands where the heat is greater, or the climate is better suited to the rapid and luxuriant growth of plants than our own is, one often observes how largely the ordinary forest trees are made the ladders

G

up which the climbers swiftly run, in some instances con-
tenting themselves with the mere production of a leafless
stem until the very summit of the tree is reached, when a
perfect forest of leaves and flowers is rapidly shot forth, in
which the plant seems to revel and glory. All who have
travelled in the East, or visited the more luxuriant forests
and botanical hunting grounds of South America and the
West Indies, will have observed this ; and many are the
references which are to be found in the works of travellers
to this striking peculiarity as first witnessed by them.
The late Captain Gill was much interested, as he passed
through Burmah, in witnessing the triumphant way in which
a vast creeper would encircle a graceful tree and coil
around its stem like a serpent till it reached its top ; then,
having reached the day-light, begin to put forth its foliage
and blossoms, while it laughed to scorn its supporter, which
had been gradually strangled to death by its ungrateful
sycophant ! In these tropical and equatorial forests, the
" Bush-ropes and other climbing plants wrap round the
largest of tree-trunks, twist themselves in and about the
arboreal foliage, and eventually reach the highest points,
reserving their vegetative power until then, and putting it
forth under the most favourable circumstances as to light
and heat. In the Central American forests, among such
successful schemers may be mentioned *Marcgravia umbel-
lata,* which flattens and moulds its own stem around the

trunks of more robust forest trees, puts forth root claspers to embrace them, and so raises itself, like a *parvenu*, above those which help it. And eventually, when it has reached the light above, and over-topped the foliage of the trees it climbed by, it throws out branches with ordinary round stems and leaves like other plants—just as if it had not cheated, and strangled, and done all kinds of vegetable crimes before it satisfied its own ends!

" Wallace mentions one of the most extraordinary of the Bauhinias he saw in the forests of the Amazon's valley, which had a broad flattened stem, that twisted in and out in the most singular manner, mounted to the tops of the tallest forest trees, and thence hung down in gigantic festoons many hundreds of feet in length."* The Marcgravia and its allied rival the Norantea are very remarkable plants; they are not only handsome and clever climbers, but are possessed of curious pitcher or hood shaped bracteæ, which somewhat resemble the vessels formed by the metamorphised Leafstalks and leaves of the so called Pitcher-plants (*Nepenthes*), and on the same plan as the remarkable trap arrangements found in the Sundew (*Drosera*) and Venus' Fly-trap (*Dionæa*).

A further illustration of our subject may be extracted from Mr. Bates' interesting work entitled, " The Natural-

---

* " The Sagacity and Morality of Plants," by Dr. J. E. Taylor, pp. 47-8. " Freaks and Marvels of Plant Life," chapter ix.

ist on the Amazons." The writer says (p. 17), "Below, the tree trunks were everywhere linked together by sipos; the woody flexible stems of climbing and creeping trees whose foliage is far away above, mingled with that of the taller independent trees. Some were twisted in strands like cables; others had thick stems contorted in every variety of shape, entwining snake-like round the tree trunks, or forming gigantic loops and coils among the larger branches; others again were of zigzag shape or indented, like the steps of a staircase, sweeping from the ground to a giddy height. It interested me much afterwards to find that these climbing trees do not form any particular family. There is no distinct group of plants whose especial habit is to climb, but species of many; and the most diverse families, the bulk of whose members are not climbers, seem to have been driven by circumstances to adopt this habit. There is even a climbing genus of palms (*Desmoncus*) the species of which are called in the Tupi language *Jacitara*. These have slender, thickly-spined and flexuous stems, which twine about the taller trees from one to another, and grow to an incredible length." This note well illustrates the way in which evil habits grow and twine themselves about the person, until eventually they become perfect masters of the situation. Our own curious plant, known as Travellers' Joy or Virgin's Bower (*Clematis Vitalba*), the first to be

described in most botanical manuals as belonging to the
great Ranunculus family, is a familiar illustration. This
plant is somewhat fastidious respecting the kind of soil
on which it will deign to grow, but once established it is
not easily dislodged. I remember recently taking a
walk along a country lane in North Oxfordshire where
the Clematis was growing luxuriantly, and seeing an old
man trimming the hedge I inquired of him the name of
the plant. He replied that the people called it "Honestwood," but he did not think the name a good one, for
he could remember the time when only one single plant
was to be seen all round the neighbourhood, but now it
had spread so fast that it did not matter what the mound
(*i.e.*, hedge) might be made of, it would choke Hazel,
Maple, Thorn or anything else, and they could not keep
a tidy hedge where the Honestwood established itself.
Vices in like manner frequently bear euphonious epithets,
for it is not always wise to call a thing by its right name.
In Notts a few years ago, when prize-fighting and other
forms of gambling were indulged in, it was no unusual
thing for a man to find himself after a night's carousal
entirely relieved of all his money. When he had become
" fresh," his " friends " had " touched " his pockets !
Well might one at such times pray to be delivered from
the gentle touch of friends. Trench has forcefully dwelt
on this habit of covering over the ugliness of a thing by

giving it a name which ought to be applied only to things of worth and virtue.

As I have referred to the destructive qualities of the Clematis I may here remark that it possesses a most peculiar and interesting method of gaining its point, and raising its head above, and by means of, other shoulders. It does not wind its stem round and round as the Honey-suckle, Hop, or Convolvulus would do, but has the pro-perty of twisting its leaf-stalk or petiole in the direction of the stick or branch against which it may be growing. Hence, if the plant is growing amongst a bush or hedge-row formed of Hazel, the twigs of the latter plant will gradually be encircled with strong, tough leaf-stalks, which serve to keep the straggler in its place, and give it a footing higher and higher up the hedge or tree till at last it overtops the whole, and then throws out a power-ful supply of new branches, leaves and flowers, which completely swamp the undergrowth and prevent its growth. Wallace, speaking of the Virgin forests through which he travelled on the Amazon, confirms the state-ments of Mr. Bates: "Its striking characteristics (he says) were the great number and variety of the forest trees, their trunks rising frequently for sixty or eighty feet without a branch and perfectly straight; the huge Creepers which climb about them, sometimes stretching obliquely from their summits like the stays of a mast sometimes winding around their trunks like immense serpents waiting for their prey. Here, two or three together, twisting spirally round each other, form a complete living cable, as if to bind securely these monarchs of the forest: there, they form tangled festoons, and, covered themselves with smaller Creepers and parasitic plants, hide the parent stem from sight." I leave the reader to follow out the train of thought for himself.

HYDRA-HEADED MONSTERS.

HILE some plants seem to grow with as much leisureliness and ease as if they had eternity before them in which to reach perfection, others, like vices, crop up everywhere, and multiply with marvellous rapidity. They can be compared to nothing more aptly than to the Hydra, which, being possessed of nine heads, dwelt in the Lernean marshes in Argolis. Hercules being sent to slay the monster, cut off one of its heads when instantly two shot up in its place. Hence things which go on increasing the more you attempt to overcome, cut off and eradicate them, are said to be Hydra-headed. And what a host of Hydras do we find among flowers and plants ! I have already referred to the methods employed by the composite plants, such as Thistles and the Dandelion, for dispersing their seeds; now think for a moment about the wonderful arithmetical problems which the seeds of plants present. Suppose that one Thistle grew only two flowers in a year and each flower head contained twenty seeds. Out of these forty seeds which are dispersed in the year of grace, we will allow that ten are unable to find suitable places of growth, and ten others are devoured by small birds and other enemies. In the second year of grace twenty plants are growing in the place of one. These twenty next year at the same rate will produce four hundred, and in ten years we shall find that this very low computation gives us no fewer than 512,000,000,000 Thistles! Thus under favourable conditions one seed would in the course of a single decade people whole acres of land with living plants of a most pernicious and injurious kind. Well says the old proverb :—

> " One year's weed
> Seven years' seed."

Let the Groundsel go to seed in your garden, and your most careful attention for years to come will fail in entirely undoing the mischief then done. It is perfectly astonishing how prolific many plants are, and strange to say the most useless and injurious flowers appear the most productive, for " Ill weeds grow a-pace," in the garden and field, as well as in the heart. But there are other means by which plants, flowers, and weeds are produced. They do not all depend upon the perfection of

their seeds, which form the general rule for the higher
orders of plants. Secret and often complex methods are
employed by the smaller plants for reproduction, and to
these we will now give a passing glance. If we descend
to the lowest forms of plant-life, those which have to be
examined under the microscope, we shall find much in
their history to wonder at and admire. They consist of
simple cells possessed of the duplex power of coalition
and division. At certain times two of these simple structures will be drawn together when one will entirely swal-
low up the other. Having thus formed one parent cell out
of the pair, this monster now commences a process of division, and when it has split up into a pair the newly formed
couple, again divide and so on for a long time till a whole
colony of young cells or plants has been formed, whereupon
the process of conjugation and division will again be repeated, till "the little one becomes a thousand." In the
case of some of these minute forms of plant-life the newly
formed plants entirely separate from the parent, each cell
being in fact a perfect and independent plant. But in some
instances the cells hold together and thus form a string,
which supplies us with a clue to the connection between
the very lowest and highest forms of life. The individual
cells which go to form this chain, each possess the power
of dividing or forming new cells ; and so while children
and grandchildren are being begotten with surprising

rapidity, they still possess that cohesive quality which makes them look like a miniature Scottish or Chinese clan, or an original Hindû family circle. And these colonies are at the same time so Liliputian in their individual and collective capacity that thousands of cells can co-exist in a cubic inch of water! Now look at the micro-fungi to which I have already directed attention in speaking of the history of smut in corn. Where shall we find these things? Dr. Cooke will reply—"We need not travel from home for examples : the unwelcome dry-rot may have committed its ravages beneath our kitchen floor, or the walls of our cellars, and our casks or bottles of wine may be infected with numbers of this ubiquitous race. Can we find no morsel of bread or cheese upon which a mould is flourishing? No towel or other article of household linen presenting traces of mildew? Are we perfectly certain that all our preserves are unvisited? or, to come nearer to some of us, all our books untouched? But in places which many would consider more unlikely still, we may look for and expect to find fungi: on white-washed walls, plaster ceilings, dirty glass, old flannel, and old boots and shoes, or leather of any description ; on carpets, mats, and boards ; and even the plants of our herbaria must be watched against their ravages. Animals bear them about on their horns and hoofs, and the house-fly often carries on its body the vegetating fungus which ultimately deprives it of life. The yeast that is employed

in fermenting our bread and our beer is a fungus, as well
as the mildew and smut that infest our growing corn.
From cesspools and traps the minute dust-like spores of
hidden fungi rise into our dwellings ; unseen they float in
the air, entering everywhere, depositing themselves everywhere, and vegetating wherever the conditions are favourable to their development." Who has not observed
the manner in which the puff-ball has sent forth clouds
of dust when a ripe specimen has been trodden on or
squeezed ? The dust then emitted was made up of
myriads of tiny spores, which, set free from their case,
were eager to find a congenial spot on which to open
house for themselves ! Come now and look at this leaf.
It is from the well known Coltsfoot (*Tussilago Farfora*),
the farmer's pest. You observe that the leaf is marked
here and there with orange-coloured spots, as though a
painter had been swinging his dirty brush round to clean
it, and some of the drops of his priming had fallen on the
plant close by. You perhaps have not noticed these
spots before, but you will generally find abundance of
them in any field where the Coltsfoot is to be obtained.
Let us now put our leaf under the microscope in such a
way that the lens shall directly focus over one of these
spots. Now look ! Did you ever see an object more
surpassingly perfect and beautiful ? The microscope has
shown that this spot is made up of a hundred little
baskets of delicate silk-work filled with oranges. They are

in reality minute fungi bearing the botanical name of Æcidia, and this one, from its being found on Coltsfoot, is called *Æcidium tussilaginsis.* The myriads of miniature oranges which we see arranged within the open baskets below, are in reality the spores by means of which new plants will be produced, and for the beauty and simplicity of their structure they may be compared with the cells of the Protococcus, or the spores of the Uredo already described.

But it is not upon the Coltsfoot alone that this hydra-headed fungus grows. It may be found in spring time upon the leaves and stems of the Nettle in such large patches as to strike the eye of the pedestrian or equestrian as he hurries along the road; it is to be seen distorting the leaves of the pretty wood Anemone, and covering them with tiny dots from the tip to the petiole; it occurs on the Gooseberry, and has recently been found on the Lily of the Valley; while the form which occurs often on the leaves of the Barberry has proved of special interest. I recently gathered this species (*Æcidium berberidis*) at Witney, in Oxfordshire, when on my way to visit "Spurgeon's Tabernacle," near Minster Lovell, for an account of which romantic spot the reader may be referred to "Lectures to My Students" (Second Series, p. 83). Recent investigations tend to confirm the idea that there is an intimate connection between the cluster-cup found on the Barberry, and the rust (*Puccinia*) which so fre-

quently attacks the wheat stalk.* You would be surprised to learn what a difficult matter it is to destroy
these minute forms of life. Should you dry up the pond
or water-butt in which your Algæ were found growing,
and allow the scorching sun of summer to shine on the
place or cask till every sign of their presence had disappeared, yet, no sooner would the vessel or pool be re-
filled with water than your hydra-heads in multitudes
would be present everywhere. In a dry state the tiny
plants are often caught up by the wind and transported
to some distant spot, where finding sufficient moisture for
their purpose they again commence their work of reproduction. One has but to spend an hour in studies like
this in order to realise with what rapidity even the smallest
forms of life establish themselves everywhere, and thus we
obtain a vivid illustration of the properties of vice. Cut off
its head and it sends forth two : dry up its germs, and
they have only to seek a congenial spot when they at
once startle us with the rapidity with which they again commence their work ! Fire seems to be almost the only thing
that will destroy the vitality of many of these micro-plants,
and the myriad forms of vice which grow around us
seem to require a force of the same description in order to
bring them into check, and lead to their extermination.

* See, however, W. G. Smith's able work on " Diseases of Field
and Garden Crops."

HE Bible states that the whole creation is now groaning and travailing as the result of some great evil which has overtaken the world (Romans viii. 22). Elsewhere we learn that the cause of suffering and misery is sin, transgression of the Divine law ; and we are told that when man followed the advice of the serpent, and threw off the yoke which God had placed upon his shoulders, resolving to be a free man, he brought himself and others into bondage and captivity. Hence, whatever method may be adopted for interpreting the statements made respecting the origin of evil, we find that the plants and flowers supply us with abundant

EVIDENCES OF THE FALL.

We will dwell on that particular evidence to which reference is made in Genesis iii. 17-18, " Cursed is the

ground for thy sake ; in sorrow shalt thou eat of it all the
days of thy life ; Thorns also and Thistles shall it bring
forth [*or* cause to bud] to thee, and thou shalt eat the
herb of the field." I cannot better illustrate what I wish
to enforce in this connection than by quoting the beautiful and thoughtful words of Dr. Macmillan ("The Ministry
of Nature," p. 104) : " Many individuals believe that we
have in this curse the origin of Thorns and Thistles, that
they were previously altogether unknown in the economy
of nature. It is customary to picture Eden as a paradise
of immaculate loveliness, in which everything was perfect,
and all the objects of nature harmonised with the holiness
and happiness of our first parents. The ground yielded
only beautiful flowers and fruitful trees—every plant
reached the highest ideal of form, colour, and usefulness
of which it was capable. Preachers and poets in all ages
have made the most of this beautiful conception. It is
not, however, Scripture or scientific truth, but human
fancy. Nowhere in the singularly measured and reticent
account given in Genesis of man's first home, do we find
anything, if rightly interpreted, that encourages us to
form such an ideal picture of it. It was admirably
adapted to man's condition, but it was not in all respects
ideally perfect. The vegetation that came fresh from God's
hand, and bore the impress of His seal that it was all
very good, was created for death and reproduction ; for

it was called into being as 'the herb *yielding seed*, and the fruit-tree bearing fruit, whose seed is in itself.' We must remember, too, that it was before and not after the Fall that Adam was put into the garden 'to dress and keep it.' The very fact that such a process of dressing and keeping was necessary, indicates in the clearest manner that nature was not at first ideally perfect. The skill and toil of man called in, we suppose that there were luxuriant growths to be pruned, tendencies of vegetation to be checked or stimulated, weeds to be extirpated, tender flowers to be trained and nursed, and fruits to be more richly developed . . . From this law of the universal diffusion of plants arises also, of necessity, the tendency to form thorns. For when plants are struggling with each other for the possession of the soil, some species must be so crowded that they cannot develop themselves freely ; and therefore, owing to the exhaustion of the soil and the pressure around them, they must pro-duce abortive Branches or Thorns. We have every reason to believe that this law existed in the pre-Adamite world, and was in full operation before the Fall . . . The Hebrew form of the curse implies, not that a new thing should happen, but that an old thing should be intensified and exhibited in new relations. Just as the rainbow, which was formerly a mere natural phenomenon, became after the Flood the symbol of the great world covenant ;

just as death, which, during all the long ages of geology, had been a mere phase of life, the termination of exist-ence, became after the Fall the most bitter and poisonous fruit of sin ; so Thorns which in the innocent Eden were the effects of a law of vegetation, became significant in-timations of man's deteriorated condition." It is not, however, alone in connection with Eden and the Fall of Man that we find Thorns and Briers spoken of as evidenc-ing the degradation of the human race. In various in-stances prophetic Scripture employs the figure. Isaiah speaks of the desolation which shall befall the enemies of the church in the following expressive terms (ch. xxxiv. 13), "Thorns shall come in her palaces, Nettles and Brambles in the fortresses thereof; and it shall be an habitation of dragons, and a court for owls." Elsewhere (ch. xxxii. 13), we read that "Upon the land of my people shall come up Thorns and Briers ; yea, upon all the houses of joy in the joyous city; because the palaces shall be forsaken ; the multitude of the city shall be left ; the forts and towers shall be for dens forever, a joy of wild asses, a pasture of flocks ; until the spirit be poured upon us from on high, and the wilderness be a fruitful field, and the fruitful field be counted for a forest." Here we are reminded that whereas Briers and Thorns, Brambles, Nettles and Weeds follow the curse, luxuriance and beauty crown the land, which secures the blessing of heaven. "The wilderness and the solitary place shall be glad for them, and the desert shall rejoice and blossom as the Rose. It shall blossom abundantly, and rejoice even with joy and singing; the glory of Lebanon shall be given unto it, the excellency of Carmel and Sharon, they shall see the glory of the Lord, and the excellency of our God" (Isaiah xxxv. 1-2).

In the extract given above a hint has already been supplied respecting the origin of the Thorns or prickly points with which many plants are endowed. It is, however, very doubtful whether that explanation will by any means cover all the difficulties suggested by the study of these apparently abnormal branches. Their form, shape, appearance, and use, are very varied, and Thorns or Prickles are not confined to any one order or family of plants. . Let us examine some of the more prominent and well known representatives. We need not dwell upon the Nettle, either as we know it in England, or as it is met with in the East. The Thistle family will at once recur to every mind, and if ever a group of plants was suggestive of a curse this is. Look at the rapidity with which it spreads, think how difficult of extirpation when once established, and reflect on the almost utter uselessness of the land which is infected, or the crops among which this enemy of cultivation grows. Its very name means "the tearer," because it scratches and tears the flesh, or other material which touches it. Thistles grow everywhere, and since they could not be overlooked they have gathered around themselves a number of fancies and associations of various kinds. Respecting one Thistle with spotted leaves it is affirmed that drops of the Virgin's milk produced the stains. The Carline Thistle was employed, if tradition may be trusted, by Charlemagne for the purpose of curing the plague which was rapidly carrying off his army, an angel having mercifully pointed out the remedy. As the flower, which is large and handsome, regularly closes before rain, it is frequently employed on the continent for indicating changes in the weather, just as Sea-weed is used by our own peasantry. The Scotch Thistle has secured for itself lasting memory and admiration from the fact (if fact it be) that its thorns pierced the foot of a Dane, and so led to the delivery of the noble sons of the soil from the perils of an invasion.

> " Proud Thistle ! emblem dear to Scotland's sons,
> Begirt with threatening points, strong in defence,
> Unwilling to assault."

The Holly, too, is provided with thorns, but it may not be everyone who has observed that the uppermost leaves of a Holly bush are devoid of these protecting points. Southey has called attention to this fact in some beautiful lines which we may here take the liberty of quoting.

> "O, reader ! hast thou ever stood to see
> The Holly tree ?

The eye that contemplates it well perceives
　　Its glossy leaves,
Ordered by an Intelligence so wise
As might confound an atheist's sophistries.

" Below a circling fence its leaves are seen
　　Wrinkled and keen ;
No grazing cattle through their prickly round
　　Can reach to wound
But, as they grow where nothing is to fear,
Smooth and unarm'd the pointless leaves appear."

Then there is the Knee-holly or Butcher's Broom (*Ruscus aculeatus*), which is one of the most curious and interesting of English plants. Its true leaves are very despicable organs, but in their place we have another organ developed, which, in addition to the sharp prickle which surmounts it, bears on its under surface a flower which is in due course followed by a large Red Berry. We are here reminded of the well known Cactus tribe. If in the last-named plant the footstalk of the leaf takes the place of the true leaf, in the Cactus the stem frequently assumes the shape, and performs the functions of leaves, the true leaves having been transformed into prickles of a most defiant character. In very many foreign lands we find the prickles of such like plants turned to good account. In the Holy Land, Egypt, India, China, the West Indies, Africa, and elsewhere, the Cactus, Prickly Pear, or some allied plant, is employed for making hedges; just as the Holly or Thorn may be

at home.   I have myself seen hedges in Eastern lands
formed by the Euphorbia, and we are told that when our
soldiers have been obliged to force their way through such
protective growths, they have suffered severely from the
results.   When we bear in mind that great evaporation
takes place through the leaves, we see how admirably the
Cacti are adapted for luxuriant growth in the hottest and
most barren spots, such as rocks, deserts, the roofs of
houses, and the tops of walls.   One species of this family
is famous for the peculiar kind of prickles which it produces, on which account it is called the Tooth-pick
Cactus.   It is popularly asserted that the monkeys of
Chili were so disgusted with the prickliness of the
Araucaria, or "Monkey Puzzle" tree, and other plants,
that they have entirely deserted the place.

In our own island the family called Rosaceæ, is
specially famous for the number of Spine or Thorn
bearing plants it contains.   It is also curious to observe
that while the Medlar in a cultivated state is free from
Thorns, it has them when found wild.   This will remind
us of a similar fact connected with other trees.   The
Apple is thornless, but the Crab has very strong spikes
at the tips of many of its branches.   The Plum tree is
often quite free from spines, but those of the wild Plum,
Sloe, or Blackthorn, are possessed of surprising strength.
The Rose itself is so famous for its prickles that we often

say there is never a Rose without a thorn. This is not literally correct, although as a popular statement of a general fact it may well be allowed to pass unchallenged. The Bramble, which also belongs to the Rose tribe, is well supplied with thorns, and many a poor sheep has had his wool rudely torn from its back by the curved prickles of this well-known plant. Thorns are of great variety of shape and form. Some are straight, polished, gradually tapered, finely pointed ; others are blunt and coarse, as witness the Cactus on the one hand and the Blackthorn on the other. Some are curved so as to look like a bird's claw, and when a thing has once been caught it is not easily disentangled. Some are comparatively harmless, others are equally poisonous. Some are woody and will penetrate hard material, others are soft and hairy, only entering the delicate flesh of the unprotected animal. In the economy of nature they each have their place. While they can be turned by human art to man's account, and be made useful as defences, they also act as defences to the plants themselves. Without these bristling bayonets the young plants would often fare badly in the struggle for existence. Dr. Taylor remarks in his own felicitous style, " The possession of stinging cells, like those of the Nettle, is evidently intended to be defensive. It is only so, however, against certain enemies, as for instance, herbiverous animals ; for our Nettles are sup-

porters of many insect larvæ. The stiffened kinds of
hairs we call Thistles, and which are so abundantly
present in some species of those composite plants which
on that account receive their well known name, are
doubtless protective against herbiverous animals, although
the donkey has a mouth hard enough to be unaffected by
them. These Thistles, covering stem and even leaf, are
more abundant in some species of *Carduum* than others.
Not only must they be protective against larger animals,
but also against slugs and snails, which find it unpleasant
to trail their molluscous bodies over such a *chevaux
de frise.* The latter creatures appear to be much more
deterred by the spiny mechanism which plants possess,
than by the poisons they secrete. The edges of the
leaves in the Holly curl and stiffen into spines in such a
defensive manner, that its admirable adaptation suggested
one of Southey's best odes. In the Gorse or Furze the
leaves harden into spines, which are defensive against all
comers, although sheep feed greedily upon them when
they are young, succulent and tender. In the Cactus the
fleshy stem performs leaf functions, whilst the true leaves
are aborted into the numerous needle-like prickles which
cover its surface, and effectively protect it against the
animals to whom its abundant juices would be so welcome
in the arid deserts where the Cactus most loves to grow."
Even the Holy Land is not free from the effects of the

curse. Thomson tells us that the Thorns, and especially that kind called *hellan*, which covers the country, are so folded together as to be utterly inseparable, and being united by thousands of small intertwining branches, when the torch is applied they flash and flame instantly, like stubble fully dry (compare Nahum i. 10). Thorns and Briers, he adds, grow so luxuriantly here that they must always be burned off before the plough can operate. Hence in his beautiful Parable of the Sower, the Saviour ells us how some seed fell among thorns, and this helps us to understand the imprecation of Job when he says : "If I have eaten the fruits thereof without money, or have caused the owners thereof to lose their life, let thistles grow instead of wheat, and cockles (or noisome weeds) instead of barley " (Job xxxi. 39-40). The writer of the Epistle to the Hebrews speaks of the earth in its twofold aspect of fertility and barrenness, and says, " That which beareth thorns and briers is rejected and is nigh unto cursing, whose end is to be burned " (Heb. vi. 8). Throughout the sacred writings, in fact, Thorns and Briers are associated with the curse of God, this in its turn being the reward of sin. Thus in speaking of the vineyard of the Lord of Hosts (Isaiah v. 6.) which brought forth no fruit save wild grapes, the husbandman says, " I will lay it waste; it shall not be pruned nor digged; but there shall come up Briers and Thorns." In Hosea too we read that "The high places of Aven, the sin of Israel, shall be destroyed; the Thorn and Thistle shall come upon their altars" (Hosea x. 8); while the same prophet, foretelling the destruction of their temples and goodly places, says, "Nettles shall possess them; thorns shall be in their tabernacles." Everyone has observed with what readiness Nettles and Thorns usually grow over places which were once teaming with life and beauty, but have through some cause or other been deserted. The Hebrew word Dardar, which is employed in Genesis iii. 18, is generally derived from a root meaning "to tear" (*dûr*), so that the etymology corresponds with that of our words Thorn and Thistle. Whatever may have been the early condition of the earth, certain it is that to-day it continues to bring forth Thorns and Thistles, so that it is only by the sweat of his brow that man can eat the produce of the soil.

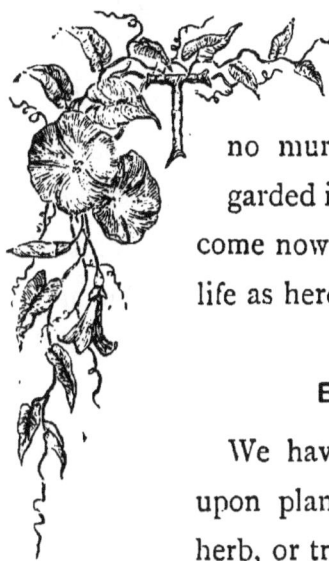

HE law says : "Thou shalt not steal," and "Thou shalt do no murder," yet these laws are disregarded in the plant world. We therefore come now to study another of the evils of life as here illustrated by

BREAKING THE LAW.

We have all noticed how insects prey upon plants, and how scarcely a flower, herb, or tree is free from the depredations of aphides, beetles, caterpillars, cicadas, or some other thievish or murderous assailant of plant life. But it is only within recent years that people generally have been made aware of the amount of evil which is wrought among flowers, grasses, herbs, and fruits by means of various species of microscopic fungi. Some of the more remarkable parasites were at one time regarded as belonging to some stage or other of animal life, but under the penetrating gaze of the powerful lens the life history of many of these enemies of plants has been traced, the parasites themselves being now defined, arranged, classified, named, just as are the flowers and plants on which they are found preying. It would not be fair if I were to convey the impression that these tiny plants do no good. They are certainly greatly needed in the economy of nature. Just as the tiny animalculæ found in putrid flesh serve as scavangers, and assist in the rapid removal of dangerous and disease-bringing portions of matter, so these insignificant fungi are equally busy breaking up the leaves of plants, and disintegrating their tissues, till eventually the decaying mass serves for food and nourishment to a new and rising race. But even the blessings of life sometimes become its greatest evils. Money is indispensable, but it is a root from which all kinds of mischief spring. Wine may nourish and stimulate, but if taken in excess it produces misery and a broken constitution. Thus with these plants of various kinds which prey upon others, while they live on exhausted and decaying matter they serve a useful end ; while they cause destruction of weeds and noxious plants which would otherwise become super-abundant their services are to be desired, but when they attack our vines, potatoes, fruit trees, corn and garden crops, we begin to think that we are having " too much of a good thing." But murder and theft are not the crimes of parasitic fungi alone. In the floral world there are many other plants which are guilty of like offences. Look at the Dodder, for example. Dr. Taylor speaks of this family of plants (*Cuscuta* or *Cassytha*) as the most ingenious of our native vegetable robbers. "There is a refinement about them not indulged in by Parasites generally : they are very particular as to the kind of plants they attack. They can only subsist, in fact, upon the sap of certain species, and this therefore restricts their parasitism ; and consequently the most abundant and widely dispersed of them is *Cuscuta Europæa*, which is least particular, and attacks Thistles, Oats, and, in short, any plants that are crowded together. When the seeds of the Dodder drop into the soil they soon germinate, and the little delicate, thread-like embryo plant makes its appearance above ground, bearing its seed covering like a protective cap at its apex. It then looks or feels about for its victim, and dies down in a few days if it cannot find one. The seeds have no cotyledons, like those possessed by the acorn or bean, but they are well provided for instead with a store of albumen (endosperm) on which the minute embryo subsists until it is fortunate enough to meet with its prey. As soon as the latter is found, the wire-like stem takes one or two coils around the victim, and develops a series of sucker-like ærial roots which penetrate into its tissues to the upflowing sap."* Thus this monstrosity is provided with all the apparatus necessary for sucking the blood and taking

---

* "The Sagacity and Morality of Plants," p. 246.

away the vital energy of its victim. The Dodder not only attacks our wild plants, Gorse, Heather, and Thistle : it preys also upon the farmers' Flax and Clover, and I have a beautiful specimen of one of the famous Banksias or Bottle-brushes from Tasmania upon which a species of Cassytha has most securely and effectually fixed itself. The woods of Hongkong, the forests of Malaya, India and New Holland, and the trees of various parts of America, are constantly the victims of this class of parasites, and in those tropical regions where vegetable life is luxuriant we can well imagine that their services are often indispensable, disagreeable as they may be to the plants they attack. In China there is a well known species of Cuscuta (*Cassytha filiformis*) which commonly lives on the Euphorbia. The flowers grow in bunches like miniature grapes, and we are told that the plant has been used medicinally in Senegal. It is curious that though these plants are of such peculiar habit and growth their flowers are similar to those of the laurel.

A very thrilling description is given by Mr. Bates ("The Naturalist on the River Amazon") of the way in which some plants commit murder. " There is one kind of parasitic tree (he says) very common near Pará which exhibits this feature in a very prominent manner. It is called the Sipo Matador, or the ' Murderer Liana.' It belongs to the Fig order. The base of its stem would be unable to bear the weight of the upper growth; it is obliged, therefore, to support itself on a tree of another species. In this it is not essentially different from other climbing plants and trees ; but the way the Matador sets about it is peculiar, and produces certainly a disagreeable impression. It springs up close to the tree on which it intends to fix itself, and the wood of its stem grows by spreading itself like a plastic mould over one side of the trunk of its supporter. It then puts forth from each side an arm-like branch, which grows rapidly, and looks as though a stream of sap were flowing, and hardening as it went. This adheres closely to the trunk of the victim, and the two arms meet on the opposite side, and blend together. These arms are put forth at somewhat regular intervals in mounting upwards, and the victim, when the strangler is full grown, becomes tightly clasped by a number of inflexible rings. These rings gradually grow larger as the murderer flourishes, rearing its crown of foliage to the sky, mingled with that of its neighbour, and in course of time they kill it by stopping the flow of sap. The strange spectacle then remains of the selfish parasite clasping in its arms the lifeless and decaying body of its victim, which had been a help to its own growth." I am afraid, after this, the story of English robbers and murderers will be tame, but such romance as this is truly saddening. We have, no doubt, all seen what efforts the

Woodbine, Convolvulus, or Ivy will make to rise in the world, but there are not many plants which are butchered outright by these ambitious climbers. The Dodder, too, is more of a thief than a murderer, and if its host can only find dinner enough for itself and its parasite, things may still go on fairly well. The Mistletoe is another of the thievish clan. It lives upon a great variety of plants —Thorn, Apple, Oak, and others—and seldom thinks of supporting itself on roots of its own. The botanist will tell you that the Mistletoe and its allies are shrubby plants—in general, true parasites, and very rarely growing in the ground. There are two species of plant which have much in common, and alike bear the name of Mistletoe—the Loranthus and Viscum. Their roots are always simple ; "and so greedily do they suck up the vital juices of the plants on which they live, that even fluids coloured by art may be detected in their transit. They will grow on almost all exogenous trees—the lactescent ones only excepted ; and in tropical America and Asia, where the more showy Loranthi are common, they often, with their pendant clusters of rich scarlet blossoms, outvie in splendour, and almost supersede, the flowers and foliage of their nursing stocks. *Loranthus Æuropæus* is, in the southern parts of Europe, a very frequent parasite on the oak, and indeed inhabits no other tree ; while the Viscum is very seldom found thereon, being chiefly

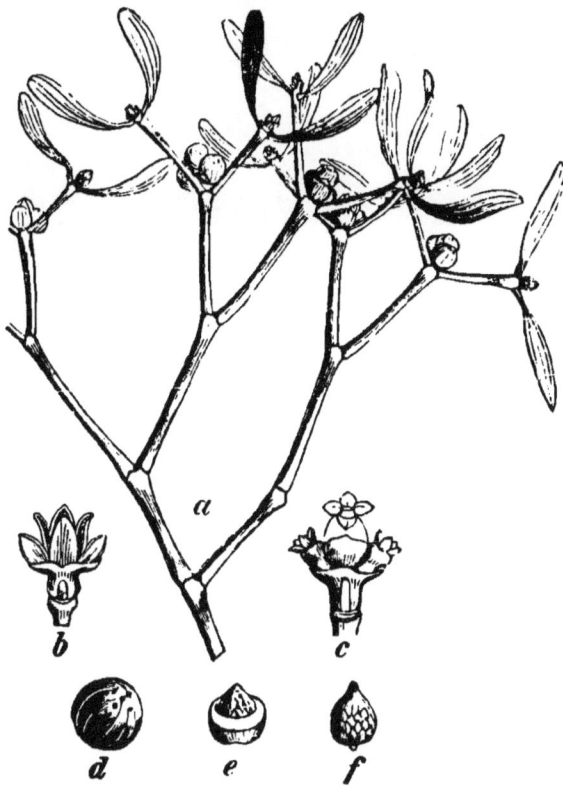

MISTLETOE. [face p. 128.

confined to the hawthorn and the apple. This circumstance has led some naturalists to suppose the Loranthus to have been the Mistletoe of the Druids, and to believe —as it is not now indigenous to Britain—that when Druidism was suppressed, every vestige of that stupendous superstition was so completely swept away that even the sacred plant was extirpated here. Such a speculation, however, seems so wild that the following is offered in its stead. The Mistletoe, although seldom found on the oak, is not exclusively a parasite of other trees, and its rarity on the former not improbably led to the preference which the old botanists, as well as the Druids, gave to the oak-mistletoe (*Viscum quercus*) over the hawthorn-mistletoe (*V. oxyacanthi*), when these vegetables were held in much repute in medicine. Hence, the very circumstance of a *search* being made for quercine mistletoe, in an age when these islands were covered with forests of oak, is opposed to the idea of the Loranthus being the plant in question, for had it then been indigenous here, the oak would have been its common, if not exclusive, habitat ; and this confirms the belief that the Viscum was the branch which the Druid went with such solemnity to cull " (" Outlines of Botany," p. 764).

It does not appear to me to be necessary to dwell long on such plants as the Broom-rape (*Orobanche*) or Toothwort (*Lathræa*), because, while they undoubtedly are

I

guilty of breaking the law, and robbing their hosts of the
nutriment they require for their own support, they are
comparatively rare, or grow on such hardy plants as are
able, in most cases, to resist the evil results of their parasitism.   It is more particularly to the ravages of what we
have termed micro-fungi that I would direct the reader's
attention, and those who would take up and follow out
this most interesting of studies may now find, ready to
hand, abundant aids.   Not only are there a number of
collectors who will readily exchange duplicates of these
objects, but Dr. M. C. Cooke has produced a delightful
handbook to the subject,* with many coloured illustrations, at an incredibly low price ; while Mr. Worthington
G. Smith has more recently done great service by publish-
ing another similar work,† which teems with facts of the
greatest possible importance and interest.   There is not
a month during the whole year when some or other of
these fungi may not be found.   You may study them
under the microscope at Christmas, by placing a little of
the green mould from a decayed orange on your slide ; or
by pulling to pieces the nuts you are eating, and examining those which are decayed ; or by looking at a piece of
green Stilton cheese.   By so doing, the spores and my-

---

* "Microscopic Fungi, Rust, Smut, Mildew, and Mould," by
M. C. Cooke, M.A., LL.D., A.L.S.   London (4th Ed.) 1878.

† "Diseases of Field and Garden Crops," by Worthington G.
Smith, F.L.S., M.A.I.   London.   1884.

celium will become familiar objects, and you will be gaining help in the manipulation of the instrument. You may find abundance of fungi of the choicest kind (*Phragmidium bulbosum*) on the blackberry leaves that are still hanging on the brambles at New Year, as I have recently done on the borders of Sherwood Forest ; while a friend has sent specimens gathered in January near Bristol. The leaves bear a violet spot on their upper surface, as a rule, when they are attacked by this fungus, and a very beautiful object do the pustules form. The spot indicates that the parasite has already commenced its ravages, and is eating out the vitals of the leaf, and so preparing the way for next year's growth. In February, you will find the pretty golden rust on the leaves of the common groundsel (*Trichobasis senecionis*), which the leaves of Anemone and lesser Celandine will begin to show signs of their pretty cluster-cups in March. Henceforth, you will see fungi regularly appearing on leaves of coltsfoot, primrose, goatsbeard, mint, willow-herb, daisy, and a hundred other plants. These are but the intimate relatives of those moulds and other fungi which attack our onions, potatoes, parsnips, and lettuce, as well as our vines, pear-trees, and barberry. What ravages these latter produce, and how they work entire destruction in the corn-field and market garden, the reader will be able to see in the books already referred to.

I may remark, in conclusion, that the plants, in many instances, retaliate and slay their enemies. A friend sent me some time since a tiny Sundew (*Drosera*) from Tasmania, the leaves of which were considerably smaller than a threepenny piece. Having cut off the tentacles from one of the leaves, I placed them under a powerful lens, when lo ! the tiny leg of an antipodean insect—too minute to be detected by the unaided eye — came to light ! This conclusively showed that the little plant of which I was in possession had slain, at least, one of its visitors. The lamented Darwin has written an elaborate work on " Insectivorous Plants," in which he gives us some wonderful information about the way in which the Sundews, Venus' Fly-trap, and other plants destroy insects. We have a few other English plants which are guilty of similar crimes, as, for example, the Butterwort (*Pinguicula*) and the Sticky Campion (*Silene anglica*), or " Catchfly," as it is called. As this little book, however, is intended to be suggestive rather than exhaustive, I will content myself with having given these few particulars, hoping that they will suffice to make the reader curious to examine the subject for himself.

BOOK III.

⸻⸱◦⸱⸻

The Ministry of Flowers

RESPECTING THE VIRTUES OF LIFE.

⸻⸱◦⸱⸻

" Dear to humanity these flowers,
The dazzling dreams of childish hours,
The hopes, the joys, the griefs of years
Have dropped on these like falling tears. "

BOOK III.

⚙️ The Virtues of Life.

N old writer says that virtue is the pursuit and exemplification of those things which experience teaches us to be best for ourselves and for society in general. This popular, though somewhat inexact and unscientific, explanation of the term exactly suits our purpose. Strictly speaking, it may be questioned whether the plants possess such things as virtues at all in the ethical sense, although the old herbalists were loud in their praises of the medical virtues of "every idle weed that grows;" but whether this point is open to dispute or not one thing is certain, *viz.*, that the plants teach many lessons, which, if properly learned and appre ciated by us, will encourage us to cultivate more largely and persistently than we have done before those habits of economy, purity, humility, liberality, perseverance and restfulness, which go a long way towards the production of a virtuous life.

We may, I think, commence this branch of our study by observing the

ECONOMY OF PLANT LIFE.

The term economy is sometimes employed with reference to a system of rules and ceremonies, sometimes in regard to regular operations, as the economy of nature ; but its most general use is that which concerns the management of affairs, the expenditure of money or material wealth, the nurturing and preservation of those qualities and materials which are of greatest importance with reference to the well-being of the person in its social or physical relations.   One man who is of delicate constitution economises his strength, and to do so refrains from speaking in public, taking part in exciting debates, keeping late hours, spending much time in society.   Another man of slender means economises his money by banking all he can spare, buying only such articles as are absolutely necessary, and adding to his little stock by further careful labours and judicious speculations.   What we thus regard as a virtue and discretion in man we may witness also in the economy of nature, whether the provision be made by the plants themselves, or be regarded as a law which has been imposed upon them, which they are bound to obey.

Begin with the root of plants, and, in order that your study may commence when the charms of the floral world are few and you have leisure for your work, take the hyacinth in your glass, or the crocus, tulip, snowdrop, among flowers properly so called, potato, yam, artichoke, onion, carrot among vegetables, and other plants of similar description famous for their bulbs or tubers. If asked why the onion or potato produced tubers or bulbs, you would probably answer, " It is a wise provision of Nature in order to secure food for man." We are so apt to think of man as the only thing worth considering by Providence that we naturally conclude that everything has reference to his well-being. But if the onion and potato possess special forms of root (speaking in a popular sense) in order to serve mankind, why should the crocus and hyacinth have similar roots or bulbs when they are unfit for human food? Evidently we have to seek another solution of the question, and the reply we give must be one which will take in all the forms with which we are familiar, whether good for animal food or not, in which the propagative principle of plants is hid in their bulbs, corms, tubers, or whatever else we may call them. The answer is a simple one : provision has been made by the thrifty plant for future contingencies, something has been laid aside against a time of need, and stores of starch and other materials have been accumulated to enable the plant to bear the exhaustion which growth, flowers, fruit and incidental demands may produce.   Biennial plants, or those which flower the second year after the seeds are sown, are subject to great strain during the period of flowering, · hence we find that in their normal condition they produce a large root, or in some other way lay in a good supply of material during the first year, against the season of flowering, when the supply will be entirely exhausted in the endeavour of the plant to secure its further propagation by means of seeds.   Take, for example, the well-known carrot or parsnip.   During the first summer after the seed is sown these plants produce enormous taproots which are filled with sugar and other nutritive matter.   Let them remain in the ground till the second spring, or take them up during the winter and plant them again when the open season returns, and this year they put forth large umbels of flowers and produce abundance of seed.   But look now at the root.   It has been in the ground as before, but the demands made upon it have been such that its radicles or rootlets have been utterly incapable of securing sufficiently large stores of nutriment to keep them supplied, and the plant has fallen back upon the stores laid up during the preceding year.   If you cut up the root now, instead of a fleshy, sweet, and useful vegetable, you find a lank, woody, and useless stock. Take the onion, and you will call to mind the fact that your bed has sometimes produced bulbs which have begun to flower the same year as the seeds were sown. But since these plants have not been able to lay by a sufficient capital, the gardener utterly ignores the seeds which may be thus produced, and takes the finest bulbs which the bed has raised for securing seed another year. When these bulbs are planted the second spring they will probably be fine fleshy bulbs some two or three inches in diameter, but when the seed-vessels are ripe and the head is seen at the top of the stalk as large as the one which formerly existed at the bottom, the gardener knows that its success has been secured at the expense of the bulb. These habits of economy have been of great service to the animal world and to man, as well as to the plants themselves. Aware of the possession of these qualities by the plants, man has cultivated those various vegetables to which we have referred for his own ends, and onions, garlic, potatoes, carrots and similar plants have come to be regarded by us as indispensable articles of diet. "The farmer, by increasing (by means of cultivation) the size of his root crops, increases the store of nourishing matter, and then by removing them from the soil in the autumn he preserves for the use of his stock nutrient substances which would otherwise have gone in the ensuing summer to support the growth of

flowers and seeds." In the case of the potato we find the
flower growing this year from the tubers produced last
year, and every novice in gardening is aware that if the
blossoms of the potato are abundant, and are allowed to
produce " potato balls " or seeds, the underground tubers
suffer more or less as the result. The common artichoke
produces underground tubers in much the same way as
does the potato, but it is not often that the plant blossoms
with us. During the month of October, 1884, I was
visiting the Isle of Wight, and whilst walking one day
near the Ventnor Hospital, saw a lad with what appeared
to be a branch containing miniature sunflowers. I at once
recognised them as the flowers of the Jerusalem artichoke,
which the boy also assured me they were. But in my
own garden at Brackley, on the borders of Northants and
Oxon, I did not obtain a single flower from the whole of
my plantation, which was a strong, healthy, and large one.
The ingredients of which the various kinds of tubers and
bulbs are composed vary greatly in different plants, according to the use they have to serve, the nature of the
plant itself, the situation in which it grows, and so on.
Artichokes and turnips are much more watery than potatoes, and we are told that the hog-plum (*Spondias tuberosa*)
holds a large quantity of clear fluid, sometimes upwards
of a pint being found in its tubercles. One kind of hog-
plum (*S. mangifera*) is known in Sanskrit as the Amrâta,

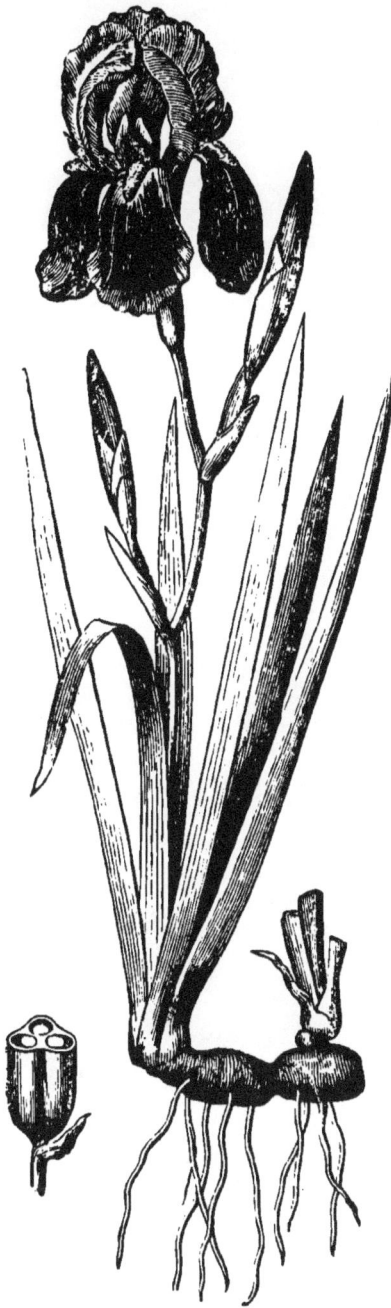

IRIS (WITH BULB). [*face p.* 140.

a name which it has received from its similarity to the mango which is called Amra. Just as a certain kind of palm which we have often seen growing in the East under the name of the Traveller's Joy (*Urania*) is made to yield the water which has collected in its trunk for the refreshment of the traveller, so Livingstone tells us that certain plants found in the Desert of Sahara are made to serve an important use among the inhabitants of that un-inviting waste, by giving up the store of water which they have accumulated in their roots or tubers. These are the veritable counterparts in the vegetable world of the camel in the animal.

Every one is familiar with our common Arum,* popu-larly known as Lords and Ladies (*Arum maculatum*), and many are aware that the root of that plant has been largely used in commerce. Nowadays the feeling in its favour is not so strong as formerly, but the fact nevertheless re-mains that the corm of the Arum contains, among other things, a great deal of starch. Now in the East, Arums are found which grow to a great size, and Dr. Hooker, in his interesting *Himalayan Journals*, tells us (Vol. II., p. 69) that he once came to a spot where some great tuberous rooted plants of this family were abundant. He adds : " The ground was covered with small pits, in which

---

* A most interesting study of this plant will be found, with il-lustrations, in *Science Gossip* for March and April, 1885.

were large wooden pestles; these are used in the prepara-
tion of food from the Arums, to which the miserable inhabitants of the valley have recourse in spring, when their
yaks (or native oxen) are calving. The roots are bruised
with the pestles and thrown into these holes with water.
Acetous fermentation commences in seven or eight days,
which is a sign that the acrid-poisonous principle is dis-
sipated; the pulpy, sour, and fibrous mass is then boiled
and eaten, its nutriment being the starch." Space will
not permit me to speak of the Yams and sweet Potatoes
which in China, the South Seas, and elsewhere take the
place of our cultivated Solanum; but these plants supply
further illustration of our subject, and show the economy
which is rigidly practised.

Let us return to give another glance at our common
wild flowers, and we will commence with the Orchids,
about the blossoms of which more may be said by-and-
bye. The most common and best known English representative of this large and interesting family of plants is
the purple Orchis, commonly known as Ramshorns (Sussex), King Fingers (Midlands), Ganfer Greggles (Somer-
set), or Cuckoo-flower. Unlike the foreign Orchids which
have recently been so much prized and sought for, our
home-grown specimens live in the soil, and if you will dig
up a specimen with care, it will be found to contain at its
base a couple of tubers ranging from the size of a filbert

to that of a horse chestnut.   One of these miniature potatoes will be found to be flabby and shrinking if the plant is well advanced, while the other is firm and white. Between them grows the stem bearing the flower.   These ophrydo-bulbs, as they have sometimes been termed, contain a store of starch and gum for the nutriment of the plant, and the flabby tuber is the one which has already begun to part with the material which, during the past year, it had been able by economy and care to lay aside. You will occasionally find the bulbous-rooted Crowfoot (*Ranunculus bulbosus*) presenting a similar peculiarity, though as a rule the exhausted bulb decays much more rapidly than does that of the Orchis.   The common Pignut (*Bunium flexuosum*), which belongs to the Carrot and Parsnip family (*Umbelliferæ*), is another case in point; and there are not many village school boys who have not dug out the sweet tubers of this plant as they have passed through copse or wood on their way to or from the seat of learning.   It is this accumulation of nutritive matter which enables bulbs to grow when placed in glasses with only a little water and the rays of sunlight to stimulate them.   The water alone would not be of any service to many species of plants, which would need a very stimulating and productive soil to enable them to grow at all. So great is the amount of material accumulated by some bulbs, that they will bear flowers for two or three years in succession before being perfectly exhausted ; but, on the other hand, the finest flowers are those of the first year ; and if a lily or other bulbous plant have its flowering spike cut off for a year or two, the flower will by-and-bye reach a greater degree of perfection than ever. So exhausting is the process of flowering, that some plants will not venture to put forth blossoms at all for several years, and gardeners generally advise you to nip off the flower buds of young and thriving trees till they shall be large enough to bring their fruit to perfection before becoming exhausted. We often hear it remarked that the Apple season may be a poor one because it was so heavy last year. Why is this ? It is a lesson in economy. The trees had accumulated material, and under the influence of a favourable season gave forth the best they possessed, even at the risk of exhaustion. Now they must again economise, or they will become bankrupt.

Turning to another view of the subject, we learn that the plant gives evidence of its study of economy by the seeds it produces. The seeds of the Pea, Chestnut, Oak, or Orange, various as are the kinds, shapes, and constituents, all bear evidence to the care which the parent has bestowed upon them in order that they may obtain a fair start in life. The Acorn consists of so large a quantity of valuable nutriment that for ages it constituted an important article of human food, and swine still enjoy feeding on it. This arises from the fact that when the Oak was providing for its offspring, it laid up within the horny covering of the Acorn a store of material which would supply the young plant with the necessary nutriment and strength when it was attempting to accommodate itself to mother Earth and grow into a tree. Here is an Acorn which I have just picked up in Sherwood Forest, and although it has not yet put forth a single root, you will find it has begun to sprout, drawing meanwhile on its own resources till the time shall come when it can secure sufficient nutriment from the soil. If you are anxious to know more about the constituent parts of the Acorn, cut it into small portions and crush it in cold water, when the latter will become quite milky. Put a drop of this water under the microscope, and you will find that it is full of granules. Now mix a little tincture of iodine with your water and it will turn blue, teaching you that starch is found in the Acorn. You can try a grain of Wheat, Indian Corn or other cereals in the same way, and you may learn from this experiment that when you are mounting sections for the microscope, a beautiful tint will be given to the cells which contain starch if the section be first immersed in a solution of iodine. Again, not only does the Hazel-bush, Oak, Chestnut-tree, or other nut-bearing plant provide its fruit with starch or other nutritious matter, further evidence is supplied of the care which is bestowed on the seed by the horny covering or shell in which it is hid, this covering being in some instances lined with soft vegetable hairs, which form a nice warm bed in which the kernel may repose during the frost and snow of winter without being destroyed by the cold. Other fruits are provided with other coverings, either as a protection from animals, or as an inducement to birds and other creatures to pluck them and so disseminate their seeds. Of this we may have more to say by-and-bye.

Plants do not, however, give evidence of economy in their roots and seeds alone. Almost any plant you may take up will afford you an illustration of our subject in one or other of its organs. Some plants are very careful as to the way in which they exhaust their vitality in petals or corolla. A plant which is to be fertilized by the wind —as all our catkin-bearers, hazel, willow, and others are —will be obliged to produce a large amount of pollen, in order that when the wind blows some of it may be carried away to the pistillate flowers on another part of the plant, or on a separate individual. Now the production of good pollen is very expensive and exhausting work, consequently wind-fertilized flowers are obliged as a rule to abstain from the production of showy flowers. A man who, with a limited income, wishes to buy a valuable horse must not spend a great amount of money on finery

for his wife and daughters, so a plant which has to produce much flour cannot display gaudy flowers. In the
same way, it will be observed that flowers which are self-
fertilized will usually have very insignificant blossoms, it
being necessary for the plant to reserve all its surplus
strength, in order to enable it to cope with flowers which,
by cross-fertilization, stand a better chance of holding
their own and establishing their position.

There are some plants which make a very lovely show
by means of their coloured petals, but in many instances
these plants have been obliged to curtail their expenses
in another direction, in order that they may meet the
demand which is thus made upon them. It will be
observed that many of our earliest flowers put forth their
blossoms before the leaves appear. I am not aware that
attention has been drawn to this peculiar and interesting
evidence of economy. Flowers which appear in
February and March, or during the winter season,
have not the tide in their favour. Hence we find the
Daphne (*Mezereon*), the Coltsfoot, the Black-thorn,
Jasmine, and many other plants bearing flowers before
the leaves; while in other instances the leaves are out of
all proportion to the flowers, being quite insignificant in
the crocus for example. It is so in the flora of other
lands as well. The well-known Japanese apple (*Pyrus
Japonica*), so frequently found trained on walls, is often

in bloom before the leaves appear; while the Chinese Tree-pæony (*Pæonia Moutan*), as I have seen it at New Year's tide in the East, often has not a single leaf when its flowers are in their prime.   The same may be said of another curious plant (*Enkianthus*), which is as great a favourite with the Celestials for decorating their houses as the holly or mistletoe is with ourselves.   This shrub bears flowers which resemble an inverted bell—which accounts for the Cantonese name of *Tiu chung Fá;* and as it grows abundantly in the south of China, large bundles of the branches are cut and brought to Canton every winter, just before New Year.   When cut, the flowers are as yet unopened; but when the branches have been allowed to stand a few days in vases supplied with water, the bunches of waxy blossoms come out to perfection, and are a great ornament on this festive occasion.

Every observer of nature is aware that certain flowers open their petals only at night, others only during the early morning, at mid-day, or in the afternoon.   The white Campion and evening Primrose are examples of the former; then come the Goat's-beard, the Hawkweed, and many others.   The night-flowering plants are usually fertilized by moths, and form an interesting study in reference to economy.   You will notice that the petals are white or pale yellow, for gaudy colours would serve no useful purpose to plants which only open shop after sun-down; but in the place of these bright petals there is usually a grateful perfume, and the flowers are large, so that they can be seen and alighted on by the moths. Now, just as the Campion and other flowers close up the corolla when the sun comes forth, that they may economise during the day, so the common flower closes its eye at night to prevent undue exhaustion. We have all noticed the way in which, one by one, the flowers go to sleep, and hence has arisen the Floral Clock. In wet weather, too, many flowers refuse to open lest their pollen or other material should be washed away and wasted. In all these methods there is clear evidence of strict economy.

Some plants economise by putting forth no branches, the stem being round and large, so that it may be stored with some valuable substance. Hence we find sago in one kind of palm, a valuable juice in another, and immense clusters of fruit in another. Other plants have found the leaf to be the most important member, in consequence of which they depend little on the flowers and seeds which in some plants are indispensable, and put forth suckers or stoles, thus propagating themselves without waste.

These are a few out of many methods adopted by the various representatives of plant life for teaching us lessons of carefulness, prudence, and economy. The ministry of the flowers on this point is worthy our attention.   " Waste not, want not," says the old proverb.   In the natural world waste is carefully provided against, and every effort is made to husband the strength of the plant and flower that its purpose in life may be accomplished; and we ought to manifest the same judicious care and foresight.

WITH all her carefulness a n d economy, nature is never niggardly. The careful can afford to be liberal. From whom do we derive the largest amount of support, the spendthrift, drunkard, and glutton, or the thrifty, careful economist? The former do little for the maintenance of our charitable and religious associations and institutions, the latter can afford to render assistance and not feel the shoe pinch. It is thus with the plants. Careful as they are they are nevertheless lavish in the bestowal of their wealth on such insects and animals as show them a favour, while the way in which they scatter their seeds broadcast over the land shows how far they are from impecuniosity. I have already dwelt on the question of

seeds, and given an illustration of what might take place
in the world if a certain Thistle were to produce a given
number of seeds which in due course should grow and
produce others.   But let us now see what actually takes
place.   Think, for example, of the Oak.   If every Oak
tree were to produce two acorns we should soon have
young trees enough to stock the Sahara.   But instead of
producing two seeds an Oak will frequently bring several
pecks to perfection, so that the people in the neighbourhood of an Oak plantation can go out and pick up
bushels of valuable nuts for the swine.   When plant-life
was scarce this property would be very valuable, but now
that the world is well stocked with useful trees the Oak
might be excused from doing so much.   We should be
satisfied with the two acorns yearly, and the right to use
the timber as we do to-day.   But then what would the
lower animals do?   The time was when man was glad
of the nutritious seed, but the Oak is willing now to feed
the swine, the squirrel, and the mouse.   If we were to
complain that the Oak provided food for the quadrupeds,
these animals in turn might point us not merely to our
Chestnut and Filbert, Walnut and Brazil, but particularly
to the splendid Cocoanut which is simply invaluable.   If
one alone in every thousand of the Cocoanuts now
grown were produced, there would be a great many
more than would be required for keeping up the

race, but when its own wants are supplied the tree does not object to give of its strength for the support of man. And see how liberality brings its own reward. The Cocoanut has fed the hungry man and quenched his thirst with its milk, provided him with a roof for his house, string for tying his articles together, fibre for brushes, mats, and other household purposes, and wood for furniture. In return for all this kindness man has taken the tree under his protection, given it the best and most suitable soil, protected its germinating seeds, cut out from its vitals the destructive worm, and quenched the prairie fire, which threatened to exterminate the race! Thus the tree is firmly established, and will live on when many another more pretentious and showy, but less liberal plant, has for ever perished from the earth. What a lesson is here. " He that hath pity on the poor lendeth to the Lord ; and that which he hath given will be paid him again."

The plants seem frequently to act on the principle, " Nothing ventured, nothing won ;" and since many thousands of seeds must annually perish from some cause or other, the flower is obliged to make liberal provision in order to meet every emergency. How often have we gathered a Poppy capsule and rattled it, while we have wondered how many seeds it contained. With what surprise have we gazed upon the seeds as they have come

tumbling out in streams, and questioned why a single
Poppy should be so extravagant. We
have perhaps walked through a piece of
rough land where the Goose Grass has
flourished, and stood amazed when we
have left it at the sight of the cleavers or
"beggar's lice," which have taken hold of
our clothes. The plant (*Galium Aparine*)
has evidently resolved to hold its own, and its numerous
seeds have consequently been provided with hooked
hairs, which lay hold of any passing object, and so get
carried away to a spot where there will be a chance of
growing. How successful this plant often is because of
the liberal supply of seed which it provides may be best
seen in early spring, when, under every hedgerow, the
tiny seedlings will be found pushing forward as rapidly as
possible that they may become established before over-
grown with grass and weeds.

It is this liberality on the part of nature which makes
it possible for man to live. The world of nature has been
made to minister to the wants of man. Hence we find
the wheat and other cereals putting forth an abundant
supply of corn, the leguminous plants supplying us
with beans and peas, the fruit trees giving us
apples, plums, currants, oranges, and a thousand
other delicacies, so that even if the supply of animal

food were cut off we should still run little or no risk of
starvation.

An interesting illustration of the lavish way in which
.some flowers dispense their gifts is to be found in the
study of the organs of fertilization. What myriads of grains
must some plants produce by way of pollen ! This branch
of hazel catkins has yielded quite a tea-spoonful, so that
a fair sized bush will produce many ounces of golden dust.
Dr. Brown remarks that in order to insure fertilisation
there is, among other provisions, a larger number of anthers and stigmas, and a superfluous quantity of pollen in
many plants. One student (Morren) found that in a single
blossom of the great flowered Cactus (*C. grandiflorus*)
there are about 500 anthers, 24 stigmas, and 30, coo
ovules. Each anther may contain about 500 grains of
pollen, so that in a single flower there may be as many
as 250, 000 pollen grains. It has been calculated by Mr.
Stephen Wilson that wheat plants produce about fifty
pounds of pollen per acre. In all other grasses, as well
as in the Coniferous and Catkin-bearing trees, there is
much more pollen than is necessary to fertilise the ovules,
even supposing that each grain took effect. Fritz Müller
calculated the number of pollen grains in a single flower
of Maxillaria to be thirty-four millions. This is not at all
incredible to one who has examined the anthers of a
flower or two under the microscope. Not only are the

grains in many instances inconceivably small, but they are packed together with such care that thousands would lie in a single case the sixteenth of an inch in length. Yet each of these grains is so perfectly finished that the most powerful lens tends rather to discover new beauties than to lay open a flaw or defect. Nor are we less indebted to the plants for the bounteous supply of fragrance and colour which they produce. There is such a thing as false economy. A good business man believes in making his trade known, and to do so he liberally supplies his customers with papers and bags bearing his name and address, puts out advertisements, and in one way or another draws the attention of the public towards himself. As a rule the most liberal business man thrives the best. So is it in the floral world. Those flowers which can command the largest advertisements in the way of petals, and can put the best and most attractive colours into their signboard, will be the most likely to have a great rush of visitors. You may easily prove this in February or March by standing near a border planted with Crocuses and Snowdrops. Let the sun come out at noonday, flooding the country with its genial rays, and soon you will hear the busy bee humming and singing as it goes about its work. The Snowdrop droops its snowy head, but the Crocus opens its golden corolla, exposes its trifid pistil and triple stamens—for in this family of plants all the organs

are arrayed in triplets—and bids the visitor enter. The
result is that soon every Crocus in the border will be
visited, and many of them will have received the attention of several guests.   Thus we learn that it would be
false economy on the part of flowers seeking the services
of insects to have poor blossoms.   But I shall be met with
an objection.   You say that the Snowdrop and Crocus
alike are raised from root organs, bulbs or corms, as the
botanist would call them.   If so, why should the crocus
have gaudy petals and sepals, while the snowdrop is content with such as are snowy white.   Here we see the
necessity of a knowledge of the history of plants, and the
country to which they belong.   You will remember that
most of our native flowers which blossom in early spring
have either white or yellow petals.   This is the case with
the Shepherd's Purse, Chickweed, Whitlow-grass, Daisy,
Snowdrop, Rue-leaved Saxifrage and others, as well as
the Celandine, Dandelion and Daffodil.   Consequently
the Snowdrop falls into the rank of those English flowers
which do not depend upon insects for their reproduction.
Independent folk can afford to do things in a quiet way,
but pushing business people must make a stir.   If the
Snowdrop never saw a bee, butterfly, or moth it would
still live on just as before, for its roots take the place of
seeds.   But the Crocus is not an English plant.   Like
many other beautiful flowers, it has come to us from foreign

shores, and its organs must naturally be studied in the
light of the climate to which it is originally adapted, and
the method which it there employs for securing its continued existence.   I am not here concerned with the his
tory of plants, but may remark that the Saffron or Crocus
is found in the East, where it has been cultivated for ages.
The Greeks called it *Krokos*, and associated with it a pretty
legend, while the Arabs called it Zafran, whence our word
Saffron.   Greece, Italy and Kashmir seem alike to have a
claim to the Crocus,* and the climate of those lands is
quite different from our own.   When we remember that
the Crocus has odoriferous stigmata which yield the saffron
of commerce, we are led to ask if there must not be some
reason for this.   Bees and insects appreciate a visit to the
plant because they are able to bring away a good quantity
of " bee-bread," and it would therefore appear that the
Crocus was supplied with a rich odour and abundance of
pollen, in order to be able to compete with the other
gaudy flowers of Eastern lands in their endeavours to secure the services of the insect world.   These details must
be borne in mind when we are dealing with cultivated or
imported flowers and plants, otherwise our conclusions
may be altogether erroneous.   The Narcissus is another
case in point.   The lovely varieties grown in China, Italy

---

* " Origin of Cultivated Plants," by M. A. De Candolle ; Kegan
Paul, Trench and Co , 1884.

and other countries are very highly perfumed, while their flowers are frequently white and waxen. The nectary is often delicately painted and so leads the insect direct to the stores of honey. The term nectary is said to have been invented by Linnæus, and refers to those organs in which the nectar or honey is secreted. Sometimes the nectary is a part of the calyx or corolla, but in the Narcissus it forms a kind of cup or crown. The odour of these flowers is very strong and possesses in some instances narcotic properties which were known to the Greeks and ancients. It was one of the flowers which was regarded with great favour for funereal purposes. While the exhalations may be injurious in confined apartments the fragrance of some varieties is exceedingly grateful. All the English plants belonging to this order (*Amaryllidaceæ*) blossom about March, and have either white or yellow flowers, but the representatives of the same Order which we find in the floras of Australia and the Orient are usually more elaborately coloured, and "got up" to attract attention.

It is in other lands, where the climate is better suited for the growth of enormous flowers and plants, that we find the best proofs of nature's liberality. In our temperate regions, as they are called—though often enough the expression is far from accurate—the biting winds and frosts of winter and early spring, the lack of

bright warm days, the attacks of blights, and the prevalence of fogs and storms, prevent the flowers rapidly
coming to perfection and making a gaudy show. But if
you could take a walk along the roads of Penang and
Singapore, visit the Islands of Ceylon or Hong-Kong, or
sail down the Straits of Malacca, you would be amazed
at the wealth, profusion, and luxuriance of the plant-world.
Imagine a single leaf of the Plantain of such dimensions
that, if it were suspended before you with its tip touching
the ground, it would not only screen your whole person
from view, but reach over your head and come down to
the ground behind you, so that if it were sewn up it
would form a sack large enough to hold a full grown
person ! Or picture to yourself that curious plant which
grows in the Malay Archipelago under the name of
Krûbût (*Rafflesia*). It is well called by the natives the
Flower of Flowers, or the Wonder of Wonders (Ambun
Ambun), and merits the epithet of the Vegetable Titan
which others have given it. Dr. Arnold shall tell us the
story of his first sight of this flower, and the account is
the more interesting because it forms part of an unfinished
letter by that ardent explorer, who died just after making
the discovery he here records. " At Pulo Lebbar [Pulo
means Island in the Malay tongue], on the Manna River,
two days' journey inland of Manna, I rejoice to tell you
I happened to meet with what I consider the greatest

prodigy of the vegetable world. I had ventured some way before the party, when one of the Malay servants came running to me, with wonder in his eyes, and said, ' Come with me, sir ; come and see a flower, very large, beautiful, wonderful ! ' I immediately went with the man about a hundred yards into the jungle, and he pointed to a flower growing close to the ground under the bushes which was truly astonishing. My first impulse was to cut it up and carry it to the hut. I therefore seized the Malay's parang (a sort of instrument like a woodman's chopping-hook), and finding that it sprang from a small root which ran horizontally (about as large as two fingers, or a little more), I soon detached it and removed it to our hut. To tell you the truth, had I been alone, and had there been no witnesses, I should, I think, have been fearful of mentioning the dimensions of this flower, so much does it exceed every flower I have ever seen or heard off; but I had Sir Stamford and Lady Raffles with me, and a Mr. Palsgrave, a respectable man, resident at Manna, who, though equally astonished with myself, yet are able to testify as to the truth.

"The whole flower was of a very thick substance, the petals and nectary being in but few places less than a quarter of an inch thick, and in some places three quarters of an inch ; the substance of it was very succulent. When I first saw it, a swarm of flies were hovering

L

over the mouth of the nectary, and apparently laying their eggs in the substance of it; it had precisely the smell of tainted beef. Now for the dimensions, which are the most astonishing part of the flower. It measured a *full yard across;* the petals, which were sub-rotund, being 12 inches from the base to the apex, and it being a foot from the insertion of the one petal to the opposite one; Sir Stamford, Lady Raffles, and myself taking immediate measures to be accurate in this respect by pinning four large sheets of paper together, and cutting them the precise size of the flower. The *nectarium,* in the opinion of us all, would hold twelve pints, and the weight of this prodigy we calculated to be fifteen pounds." We are told that it takes three months from the first appearance of the bud to the full expansion of the flower, and the flower appears but once a year, namely, at the conclusion of the rainy season. The plant takes rank by the side of the Fungi, some monstrosities of which family have been found at various times in our own land.

The leaves of such plants as the Victoria Regia, or great Water Lily of South America, are perfect marvels, and have created great surprise and excitement when viewed for the first time; but we should find it impossible to dwell on all the facts associated with plants which illustrate the unbounded liberality of nature, and as some of the matters which we might have discussed under this · heading will present themselves for examination in another place, we will now turn to another subject.

HUMILITY.

WE come from the study of the larger plants to that of humbler, yet not less lovely, forms. Here we meet with the Moss, that tiny ornament of wall and bank which is so often overlooked, but which may be made the minister of comfort to those who will read its lessons. Who has not heard of Mungo Park, the African traveller? For him the Moss had its message when lonely and sad he lay down in the wilds of the Dark Continent.

" One tiny tuft of Moss alone,
    Mantling with freshest green a stone,
        Fixed his delighted gaze ;
    Through bursting tears of joy he smiled,
    And while he raised the tendril wild
        His lips o'erflowed with praise."

Some of the Mosses are of such miniature growth that the naked eye, unless trained by long and patient search, can scarcely detect their presence, yet withal they are most elegant and curious plants. Hasselquist observed one of these small varieties (*Gymnostoma truncatulatum,* according to one author) "growing in vast abundance upon the walls of Jerusalem, and hazards a conjecture that it may be the Hyssop of Solomon. That our present Hyssopus is not the plant alluded to by Solomon there can be but little doubt. If Hasselquist's surmise should be correct, this minute 'Hyssop springing out of the wall' would contrast finely with 'the Cedar that groweth on Lebanon ;' and thus, by referring to the extremes of the vegetable world, the phrase, by a beautiful Orientalism, would comprehend the whole of which the chronicler says that the wise man spake." The Moss would thus be the Alpha, the Cedar the Omega, of Solomon's natural history pursuits. Tristram and others, however, think the graceful Caper-plant (*Capparis spinosa*) is intended.

What purpose do the Mosses serve? some will ask. The answer is easily found if we go out for a country ramble, and notice where the lowly plants exist and thrive. By their presence on rocks and mountains they serve to retain and collect the moisture which comes within their reach, and so prevent the exposed surface of these parts from becoming altogether barren and sterile. Their rootlets penetrate the rock and break up its surface into a thousand parts, while their own decaying form serves to make a fertile soil for larger growths. Here, eventually, the Fern throws out its fronds ; then the seeds of Fir and

Oak fall into the rich bed and germinate, finding sustenance enough in the thin stratum of soil to enable them to
throw out roots which gradually penetrate the rocky sub-
stratum, and thus secure a firm hold for the future tree.
"The uses of Mosses are great in the general economy
of Nature. Well have they been called, by Linnæus, her
ministers; filling up, as they do, and consolidating bogs,
clothing mountains, even to the verge of perpetual snow,
and condensing the moisture of the atmosphere; thus
giving origin to rills, and being the living fountains of
many streams." Dr. Cooke speaks, in his interesting
little book on "The Woodlands," of the impossibility of
giving, in a few pages, "any satisfactory idea of the
variety of form, the diversity of character, and yet, withal,
the harmony of design which prevail in these small and
humble plants. They are without odour to attract,
without gaudy flowers to allure, or luscious fruits to
entice, not only the pleasure seeking butterfly and the
honey-sucking bee, but even the human lounger in the
woods. They appeal not to any power of gratifying the
taste or the smell as an apology for existence, but
humbly, and without ostentation, perform the task
allotted to them in the world. Thus silently, in their
myriad forms, scattered from pole to pole—from the
burning tropics nearly to the limits of perpetual snow—
preachers are they to those whose ears are attuned to

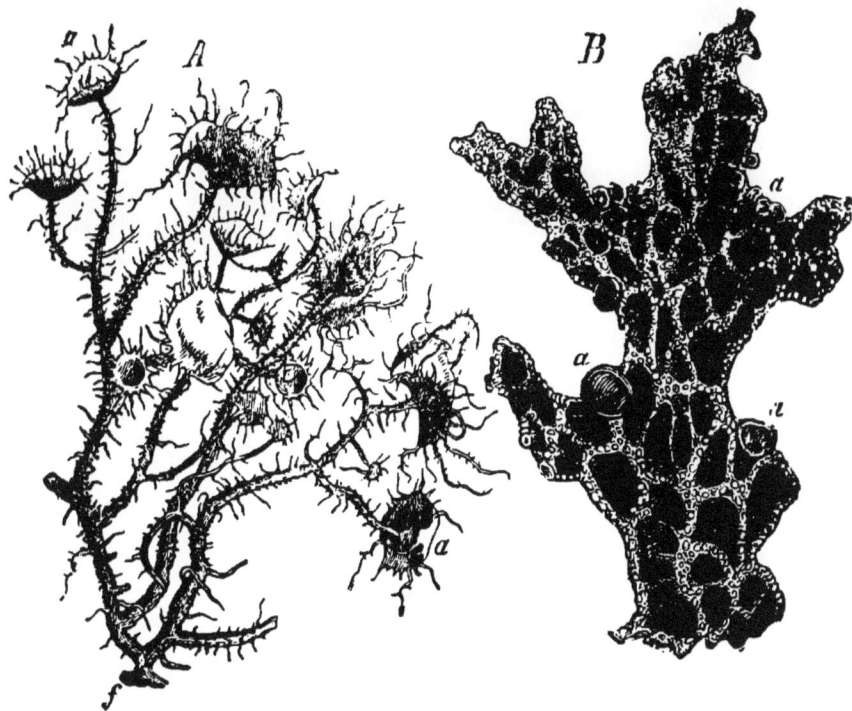

A. FRUTICOSE LICHEN.   B. FOLIACEOUS LICHEN.

[*face p.* 160.

their accents, of the wondrous power, the infinite wisdom of the great Father of all." I have just returned from a long ramble in Sherwood Forest, where I have been studying the forms and localities of the different species to be met with in such profusion there. Some grow on the open, heathery moor, others on the trees, others again on the palings and fallen timber. Here is a large patch of the thyme-leaved variety, which came from a damp place under the trees, and this, with a double set of teeth (called *Diploperistomi*), was culled from the face of a rock. Some of the flowers, to speak in a popular way, are innocent of teeth around the margin of the open capsule. By observing this you may know that the variety so marked belongs to the *Gymnostomi*, or naked-mouthed group, as Hooker designates them. Others have a single row of teeth for the peristomi (*Haplo-peristomi*), while a yet more lovely form has two rows of teeth (*Diplo-peristomi*), and looks perfectly charming under a powerful lens. When, therefore, you begin to collect Mosses, first look for the flowers, or fruits, as the organs are more generally called; and having found a ripe one, examine the edge of the capsule or theca, as the little urn is called, at the end of the stem. See whether it possess a double or single fringe, or whether it is entirely destitute of that ornament, and you have learned your first lesson in bryology. Next observe whether the fruit stalks are growing from the sides of the plant, or whether they rise perpendicularly, like a Primrose from the middle of the tuft. Some Mosses, as you have already observed, are branched ; but others grow separately like little Pine trees in a closely packed plantation, and these peculiarities will assist you in your classification. Then you may take note of the hood which is drawn over the urn. This pretty covering (termed a *Calyptra*) is sometimes very hairy, and when examined under the microscope these hairs frequently appear to be jointed something like an insect's leg. At other times it is smooth and semi-transparent. One variety is known by the calyptra slipping off entire, while in another it slits from top to bottom, and in another divides into equal portions. You will also be able to distinguish another family (*Jungermannia*) by the way in which the capsule or urn splits into four, and thus sets free its numerous spores. The spores of these plants are full of interest, so too are the less conspicuous organs which grow on separate plants and serve to keep up the growth and reproduction of the race. But as we are not here pursuing the study of practical botany, this must suffice by way of introduction to these modest plants. Mr. Step remarks that these plants should be encouraged on the fernery, " not only for the sake of their own beauty, which is great, but as helping to prevent the excessive evaporation of moisture from

the soil, and conducting the moisture from the atmosphere. They form a suitable nidus for fern spores to fall
upon, affording them the requisite conditions to induce
germination. They also serve to prepare hard soils for
more deeply-rooting plants, and for this reason are
among the most valuable of Nature's pioneers, covering
the hard rocks with a soft coating of delicate green.
Their tiny rootlets break up the surface of the rock, and
their dead bodies gradually form a thick stratum of
vegetable mould, still covered by the younger living
individuals. Here the wind-borne seeds of the giant
pines and firs find a resting-place, and germinating send
their long roots down into the fissures of the rocks for
support, and absorb their nourishment from the moss-
made mould. And in this way Nature covers up the
bare rocks with the most beautiful of mantles, that of
living greenery; and always the Mosses and Liverwort
and Lichens are the humble plants which prepare a soil
for the larger growths of Oak and Pine " ("Plant Life,"
p. 143).

Next to the mosses for humility of growth, we may
note the Lichens, which we can find, all the year round,
in almost any spot where the atmosphere is sufficiently
pure to allow them to grow. They cover the trunks of
trees, abound on palings and wood, walls, roofs of houses,
rocks, dry banks, and in almost every conceivable place.

Who has not, again and again, admired the pretty " cup-
moss?" Now, the "cups" which made this plant so

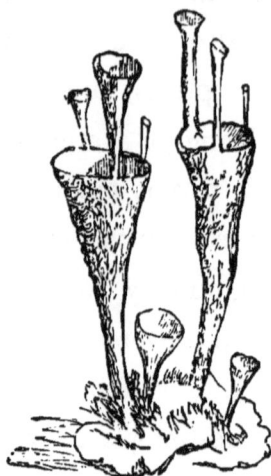

conspicuous, although it is so small, contain large numbers of little bodies like balls of dust. So, if you examine
other lichens when in fruit, you will find a similar organ
or cup, in which the spores are contained, that take the
place of seeds in the higher plants. Some of these
lichens, when perfect and full-grown, are so tiny that if
hundreds of them were not growing together, you would
not be able to detect them ; and, even when they exist in
considerable patches, you need a lens to enable you to
distinguish the various parts, and help you to discover
the cup, or apothecium. The peculiar green, rusty, grey,
or yellow colour which the bark of trees often assumes,
especially in woodland districts, is due to the presence of

minute lichens; and those patches of gold leaf, which you have so often admired on old walls, stones, and fences, are made up of a most lovely species (*Physcia parietina*). In many instances, they cannot be severed from their host, and when the collector needs specimens for his herbarium, he has need of a strong knife for wood-loving species, and a hammer and chisel for such as grow on stones and rocks. They often lie so close to the bark of a tree that only a practised eye would ever suspect their presence ; at other times, they grow into somewhat conspicuous and lovely plants. The position in which they grow is such that, during dry weather, their growth is often entirely suspended; whence it arises that years are, in some instances, required for bringing them to perfection. Some varieties are well known by all young ramblers, from their mossy appearance, as they grow on old hawthorn bushes, as well as on the oak, fir, black-thorn, birch, and ash. They have acquired the name of Tree-moss, Tree-beard, and even Jupiter's-beard, and are usually found in fir plantations, where the ground is moor-like, and in woods, where the tree-growth is too thick or decaying. Poets have not disdained to observe this fact, but, as poets are not always naturalists, they sometimes fall into error. When Gray speaks of the "rude and moss-grown beech," he shows his ignorance of this branch of natural history, for, as Johnson says, "no tree is so

little, or so seldom either rude or moss-grown," the Elm,
Lime, and Sycamore being in this respect its associates.

The lichens, humble plants as they are, have nevertheless their uses. Some of the higher orders have been
found edible, and, while the reindeer lives largely on one
variety—known as the Reindeer Moss (*Claydonia rangi-
ferina*)—man has employed another kind, called Iceland
Moss (*Cetraria islandica*), for food. Some species have
been used medicinally, and there is a variety (*Everina
prunastri*) very common in Great Britain, which used to
be much in request as an ingredient in sweet-pots and
perfumed cushions, on account of its peculiar power of
imbibing and retaining odours. Evelyn says that this
moss of the Oak " composes the choicest cypress powder,
which is esteemed good for the head ; but impostors
familiarly vend other mosses under that name, as they do
the fungi for the true agaric, to the great scandal of
physic." In these days, we generally regard such things
as out of date, and, as the spenge has taken the place of
most other absorbants, so new remedies have taken the
place of the old medicines. One or two varieties are said
to be poisonous ; others have been found to yield a valuable dye ; and in Scotland the peasantry of former times
would frequently earn considerable sums of money by
collecting dye-producing lichens from their moors and
rocks. Thus, even these humble plants have their place

in creation, and are appointed their work. "In the
arctic regions, and especially in Lapland, the reindeer
moss grows in the utmost profusion, and overspreads, as
with a coverlet of snow, plains hundreds of miles in extent. These, which to a stranger or a traveller arrived
from—what prejudice would call—a happier land, might
seem dry and barren wastes, are the very fertile fields of
the Laplanders ; for when the cold of winter has withered
up every sort of herbage, and its storms have driven man
and beast to the valleys and the woods, this lichen or
moss becomes the principal aliment of the herds of reindeer, in which consists all the wealth, and on which depends the very existence, of the natives. Thus, things
which are often deemed the most insignificant and con-
temptible by ignorant men, are, by the good providence
of God, made the means of the greatest blessings to His
creatures " ("Outlines of Botany," p. 156). Another
service is performed by certain species of lichens. If you
observe the dead branches of certain trees lying in a
copse or forest, you will probably find that they are being
broken up, and formed into tinder by means of various
agencies. Sometimes insects are at work, fungi have
attacked the ruin, or mosses have made it their abode ;
but at other times, curious forms of the ubiquitous lichen
will be found located between the bark and the wood,
loosening the former till it falls off spontaneously, or

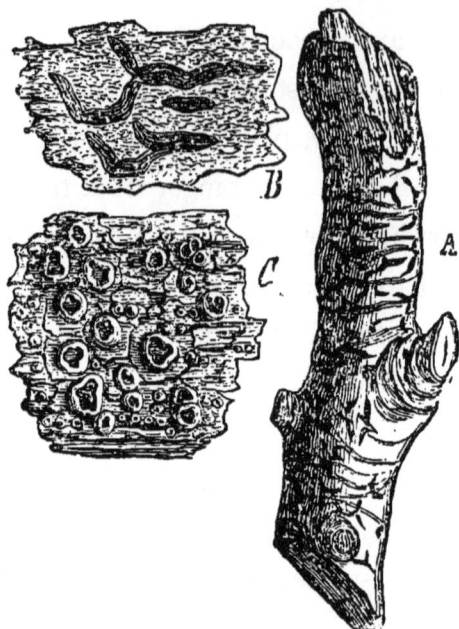

A Graphis elegans on bark of holly, natural size. *B* Slightly magnified.
C Pertusaria Wulfein also slightly magnified.

yields to the slightest force, and so leaves the bare wood
to be quickly decomposed by animal, fungoid, or atmospheric forces. How wonderful, then, are the arrangements made for the equilibrium of the natural world, and
how admirably do even the humblest, most despised, and
least known of nature's servants do their work ! Here
they add beauty to the scene, there they quietly and un-
obtrusively work away at their task of producing new
mould for trees and plants, and setting free latent and
confined gases for the future use of other creatures. This
variety is useful as a thatch to keep out the damp from

the tree it protects ; while that prevents, during the summer drought, too great an evaporation of the sap and juices. Here a species supplies the animal world with food ; there another is useful for destroying evil properties or producing a valuable pigment. There are, doubtless, many other uses to which nature appropriates these her servants, which as yet we fail to understand ; but this is certain—they all have their work and place, and may teach us that, insignificant as we are in our own eyes and in the eyes of others, God has given us our place in the world for a definite purpose. The title of Bond-slaves, which Linnæus applied to some of the lower plants, is particularly appropriate when applied to the lichens. They are, as it were, " chained to the soil they labour to improve for the benefit of others, though from it they derive no nourishment themselves. The first conquests of life over death, the first inroads of fertility on barrenness, are made by the smaller lichens, which, as Humbolt has well observed, labour to decompose the scarified matter of volcanoes, and the smooth and naked surfaces of the sea-deserted rocks. These little plants will often obtain a footing where nothing else could be attached. So small are many that they are invisible to the naked eye, and the decay of these, when they have flourished and passed through their transient epochs of existence, is destined to form the first exuvial layer of vegetable

mould; succeeding generations give successive incre-
ments to the soil, from which, when formed, men are to
reap their harvests, and cattle to derive their food; from
which hereafter forests are designed to spring, and from
which future navies are to be supplied. But how is this
frail dust to maintain its station on the smooth and
polished rock, when vitality has ceased to exert its influence, and the structure that fixed it has decayed? This
is a point which has been too generally overlooked, and
yet which is the most wonderful provision of all. The
plant, when dying, digs for itself a grave—sculptures in
the solid rock a sepulchre in which its dust may rest.
For chemistry informs us that, not only do these lichens
consist in part of gummy matter—which causes their
particles to stick together—but that they likewise form,
when living, a considerable quantity of oxalic acid; which
acid, when set free by the decay of the plants, acts upon
the rocks, and thus is a hollow formed in which the dead
matter of the lichen is deposited. Furthermore, the
acid, by combining with the limestone or other material
of the rock, will often add an important mineral ingredient to the vegetable mould; and not only this: the
moisture thus conveyed into the cracks and crevices of
rocks and stones, when frozen, rends them, and, by con-
tinual degradation, adds more and more to the forming
soil. Successive generations of these bond-slaves succes-

sively and indefatigably perform their duties, until at length, as the result of their accumulated toil, the barren breakers, or the pumice plains of a volcano, become converted into fruitful fields " ("Outlines of Botany," p. 39). Surely, then, none of us need be discouraged or despair. Crabbe has remarked that—

> " Seeds, to our eyes invisible, will find
> On the rude rock the bed that fits their kind ;
> There in the rugged soil they safely dwell
> Till showers and snows the subtle atoms swell,
> And spread the enduring foliage ; then we trace
> The freckled flower upon the flinty base.
> There all increase, till in unnoted years
> The stony tower, as grey with age, appears
> With coats of vegetation, thinly spread
> Coat above coat—the living on the dead.
> These then dissolve to dust, and make away
> For bolder foliage, nursed by their decay.
> The long-enduring ferns in time will all
> Die and depose their dust upon the wall,
> Where the winged seed may rest, till many a flower
> Shows Flora's triumph o'er the falling tower."

We leave the gentle reader to search the works of Ruskin and others, for those beautiful words in which the lessons taught by these humble plants are eloquently set forth, and would invite all who read these pages to look, during their next walk, for some of the charming objects which, in humble form, are doing such great and noble service.

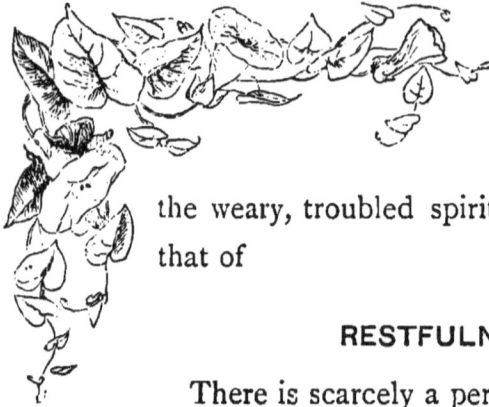

THERE is no lesson, perhaps, which the modest flower can teach the weary, troubled spirit more sweetly than that of

RESTFULNESS.

There is scarcely a person who does not at times feel chafed and sore, and the general tendency of misfortune and want of success is to make us anxious, uneasy, and fretful. When things are not so bright with us in the business as they once were, when affliction lays aside this or that member of the family, when the income is reduced or the expenses become heavier, what more natural than that we should begin to take thought for the morrow. We have already seen that economy and providence are taught us by the plants; yet the Saviour himself tells us that undue anxiety is not only useless, but injurious and dishonouring to God. His words have been already quoted, but they are so full of comfort that we may give them yet again : "Consider the lilies of the

field, how they grow ; they toil not, neither do they spin:
And yet I say unto you, that even Solomon, in all his
glory, was not arrayed like one of these. Wherefore, if
God so clothe the grass of the field, which to-day is, and
to-morrow is cast into the oven, shall he not much more
clothe you, O ye of little faith?" (Matthew vi. 28-30 :
Luke xii. 27-28).   Rabbi Simeon said on one occasion,
" Hast thou all thy life-long seen a beast or a bird which
has a trade ? Still they are nourished, and that without
anxious care.   And if they, who are created only to serve
me, shall not I expect to be nourished without anxious
care, who am created to serve my Maker ?   Only that if
I have been evil in my deeds, I forfeit my support."
There is one thing about the simple flowers of our fields
and hedgerows which always charms us, *viz.*, their bright
and cheerful, yet unobtrusive adorning.   The child just
out of school delights to pluck them as she wanders
homeward, and make them into a wreath for her brother's
neck.   The mother rejoices to receive them from the tiny
hand of the little one who has been scouring the bank
and mead, and exalts them to a place of honour in th
daintiest vase on side-board or mantel-piece.   The sick
one gazes on them with delight as they stand by the bed-
side in their purity, or scent the room with their fragrance.
Everywhere flowers are prized.   In England we gather
the Primrose, Violet, or Daisy, and carry it home with

joy; in China, the Flowery Land, the Jasmine, Rose, Lily, or Chrysanthemum will be seen carried through the streets of the cities and towns, to be sold to the poorest as well as to the most wealthy of the people, to be placed in their hair as an ornament, or on the table for the sake of their sweet perfume. There, as here, the plants are kept in pots and tenderly nurtured, for a charm seems to linger about the very name of a flower. Look at the flower of the field. You pluck the tiny Daisy whose pink-tipped blossoms bespangle the meadow in spring, but you do not need to paint it in order to make it bright and pretty enough for your nicely arranged table. It does not look shabby by the side of your costliest and choicest paintings and china, but its sweetly shaded tints tend to brighten the dullness of its surroundings. A nosegay seems to put new life and beauty into the dwelling-room.

You pluck a leaf, or gather a blade of grass, and its tints put the painter's skill to the blush, so well and highly finished are they in point of colour, as well as in shape and contour. The flower which we tread under our feet surpasses in beauty of dress and ornament the decorations of kings. "Solomon in all his glory was not arrayed like one of these." The glory of Solomon was of a surpassing kind. It had become proverbial. Even in his own days the Queen of Sheba had heard of him

and came to see if it were true, but she found that the
half had not been told her ; and when in later years men
wished to set forth the glory of a thing they compared it
with that of Solomon.   As with Dives so with this great
king, he was clothed in purple and fine linen and fared
sumptuously every day.   The dress of a man in the East
is regarded as a very important thing, and in all ages and
countries, position, rank, influence, wealth, have been indicated by means of dress.   When the people mocked
the Saviour they put on him a purple robe (the dress of
royalty) and said, " Hail, King of the Jews." We are
told that the " privileged Greeks may put on robes of
any dye except green."   In the East, the people love gay
colours, and use red, blue, green, orange, and other dyes;
but the Imperial yellow is the privileged colour of the
Emperor of China.   Though Solomon added to the
glory of his dress that of his associations and surroundings, had lovely palaces, and attendant ministers, and all
the paraphernalia of an Oriental monarch, yet the Great
Teacher says he was surpassed by the Lily of the field.
In the East, the Lily is a favourite flower, and its kinds
are very various.   There is first of all the Lotus, con-
sidered sacred by the Hindus, and especially by the
Buddhists.   It is thought by some writers that this is the
flower whose fruit was so esteemed by the Ancients, that
to be able always to eat it was worth leaving one's

country for ever.   But confusion has arisen in the minds
, of many writers through the term being applied to different kinds of plants.   We have an English Lotus, but it
is a member of the clover family and not a Lily at all.
The true Lotus is a Water Lily, occupying a middle position, for size, between the Victoria Regia on the one
hand, and our own water Lily on the other.   Passing
through Singapore on my way home to England some
years ago, I saw the ponds in the Botanic Gardens gorgeous with the pink and white blossoms of the Lily.   In
China I have seen, both within the precincts of Buddhist
Temples and in the gardens of the gentry or the agricul-
turist, large pots set apart for their cultivation; and so
much are they prized, that people who have not an inch
of garden will fill large jars with dirt and water and grow
the Lotus by their door.   When in bloom, they present to
the eye one of the most pleasing and picturesque of scenes.
The flowers are plucked and placed in front of the altars in
the temples, as offerings well pleasing to the gods.   In
Japan and India, as well as in China, the Lotus is regarded as sacred, or let us rather say, as specially pleasing
to their deities.   Dr. Gray, who resided in Canton during
my own stay in that city, aptly says in his valuable work
on China : " There are very extensive Water Lily or
Lotus ponds in the vicinity of the cities and villages of
the southern provinces.   In the western district of

Canton, such ponds are also numerous. The Water Lily, which I apprehend is the Shushan of the Scriptures, is regarded by the Chinese as a sacred plant. It flourishes during the months of July and August; and when, in consequence of the latter rain and high tides, the Canton river during these months overflows the adjacent lands, its large tulip-like flowers—some of a bright red, others of a milk-white colour, and not a few combining the red and the white—may be seen raised, as if in triumph, above the surface of the swollen waters. With these flowers the Chinese decorate their houses. The leaves of the plant are also used by shopkeepers—grocers especially—instead of paper, to wrap their customers' purchases in. The seeds of the Lotus, which are almost as large as Filberts, are boiled and eaten. From the beds of the ponds the Chinese also gather the roots of the plant, which is of an elongated form (a rhizome), and in colour like a turnip. When opened, the root, which consists of a variety of cells, has somewhat the appearance of a honeycomb. The Lotus of China is, I apprehend, of the same species as that of Egypt, of which Herodotus wrote (ii. 92) that 'so soon as the waters have reached their culminating point, there is to be seen above the surface a large quantity of the Lily species, which, by the Egyptians, are termed the Lotus.' It would appear that the Egyptians were in the habit of eating the seeds of this plant, which

they boiled and made into a paste, and then baked as
bread." The goddess Kwan Yin is represented by the
Buddhists as seated on a Lotus, while one of the sacred
invocations of this sect—"Om mani padme hum!"—
means "Hail, Jewel in the Lotus!" This is repeated
108 times by the priests in their prayers in order to
propitiate their great deity. But it would require a
volume to give all the lore attaching to this plant, and
as it evidently was not a Water Lily to which the Saviour
referred we may pass on.   Christ speaks of the "Lilies
of the field," and then says "if God so clothes the *grass*,"
&c.   Alford remarks that "these Lilies have been supposed to be the *Crown Imperial* (*Fritillaria imperialis*,
German, Kaiserkrone), which grows wild in Palestine,
or the Yellow Lily (*Amaryllis lutea*), whose golden
liliaceous flowers cover the autumnal fields of the
Levant." The Greek word (κρινον) which is translated
"Lily," was probably applied originally to a white flower,
which, on account of its innocence and beauty, was called
"The Flower" (ανθος) *par excellence*, and represented
dignity in the ancient language of flowers.   In the
Hebrew it was called Shusan, a name which is generally
derived from the word Shesh ("six"), because of the
hexaplous perianth it possessed.   The flower is spoken
of in the Song of Solomon (ii. 1), "I am the Rose of
Sharon, and the Lily of the Valleys.   As a lily among

thorns, so is my love among the daughters." In verse
16 also we read, " My beloved is mine and I am his ; he
feedeth among the lilies." A prophecy relating to Israel
says, " He shall grow [margin, ' blossom,'] as the lily, and
strike forth his roots like Lebanon " (Hosea, xiv. 5), on
which Henderson remarks that " Lilies abound in Palestine, even apart from cultivation. There are two kinds :
the common Lily, which is perfectly white, consisting of
six leaves (petals), opening like bells; and what the
Syrians call Shushan Malcha, or the Royal Lily, the stem
of which is about the size of a finger in thickness, and
which grows to the height of three and four feet, spreading its flowers in the most beautiful and engaging
manner." My friend, the Rev. W. Houghton, thus
writes respecting the Lily of the Bible : " The Hebrew
word is rendered 'Rose' in the Chaldee Targum, and by
Maimonides and other Rabbinical writers, with the excep-
tion of Kimchi and Ben Melech, who translated it by
Violet (1 Kings vii. 19). But *Krinon* (κρίνον) or "Lily."
is the uniform rendering of the Hebrew Shûshân or
Shôshannâh in the Septuagint, and is in all probability
the true one ; as it is supported by the analogy of
the Arabic and Persian Susan, which has the same
meaning to this day, and by the existence of the
same word in Syria and Coptic. But although there
is little doubt that the word denotes some plant of the

Lily species, it is by no means certain what individual of this class it specially designates. Father Souciet laboured to prove that the Lily of Scripture is the Crown Imperial. But there is no proof that it was at any time common in Palestine. Dioscorides bears witness to the beauty of the Lilies of Syria and Pisidia, from which the best perfume was made. If the Shûshan of the Old Testament and the Krinon of the Sermon on the Mount be identical, which there seems no reason to doubt, the plant designated by these terms must have been a conspicuous object on the shores of the Lake of Gennesaret; it must have flourished in the deep, broad valleys of Palestine, among the thorny shrubs and pastures of the desert, and must have been remarkable for its rapid and luxuriant growth. That its flowers were brilliant in colour would seem to be indicated in Mathew vi. 28, where it is compared with the gorgeous robes of Solomon; and that this colour was scarlet or purple is implied in the Song of Solomon (v. 13). There appears to be no species of Lily which so completely answers all these requirements as the Scarlet *Martagon* (*Lilium Chalcedonicum*) which grows in profusion in the Levant. But direct evidence on the point is still to be desired from the observation of travellers. Other plants have been identified with the Shûshan. Gesenius derives the word from a root signifying "to be white,"

and it has been inferred that the Shûshan is the White
Lily. Dr. Royle identified the Lily of the Canticles
with the Lotus of Egypt, in spite of the many allusions
to "feeding among the lilies." The purple flowers
of the *Khob*, or wild Artichoke, which abounds in the
plains north of Tabor and in the valley of Esdraelon,
have been thought by some to be the "Lilies of the
field." A recent traveller mentions a plant with lilac
flowers like the Hyacinth, and called by the Arabs
usweih, which he considers to be the species denominated
Lily in Scripture. In his excellent work entitled "Sinai
and Palestine," the lamented Dean Stanley says: "In the
spring the hills and valleys are covered with thin grass and
the aromatic shrubs, which clothe more or less almost the
whole of Syria and Arabia. But they also glow with
what is peculiar to Palestine, a profusion of wild flowers,
Daisies, the white flower called the Star of Bethlehem;
but especially with a blaze of scarlet flowers of all kinds,
chiefly *Anemones*, Wild Tulips and Poppies. Of all the
ordinary aspects of the country, this blaze of scarlet
colour is perhaps the most peculiar; and, to those who
first enter the Holy Land, it is no wonder that it has
suggested the touching and significant name of 'The
Saviour's Blood-drops.' It is this contrast between the
brilliant colours of the flowers and the sober hue of the
rest of the landscape, that gives force to the words—

' Consider the lilies of the field. . . For I say unto
you that Solomon in all his glory was not arrayed like
one of these.' Whatever was the special flower designated by the Lily of the field, the rest of the passage
indicates that it was of the gorgeous hues which might be
compared to the robes of the great king. The same
remark applies, though in a less degree, to the frequent
mention of the same flower in Canticles. . . . The
only Lilies which I saw in Palestine in the months of
March and April, were large yellow Water Lilies, in the
clear spring of 'Ain-el-Mellâhah, near the Lake of
Merom. But if, as is probable, the name may include
the numerous flowers of the Tulip or Amaryllis kind,
which appear in the early summer or the autumn of
Palestine, the expression becomes more natural—the red
and golden hue more fitly suggesting the comparison
with the proverbial gorgeousness of the robes of Solomon."
For rich illustrations of the flora of Palestine, I must
refer the reader to Mrs. Zeller's superb volume of " Wild
Flowers of the Holy Land," while Tristram and others
must be consulted for such fuller information on the subject as he may wish to obtain.

Whatever be the special meaning of the word Lily,
whether it refers to Lotus or Martagon, Anemone or Tulip,
or whether it is to be taken, in its widest sense, as referring to flowers in general, the inference to be drawn

from the words of Christ is the same. Consider the Lilies, the prettiest and choicest of flowers, or the simplest and most humble which adorn the field : the dress with which the Heavenly Creator has clothed them is far grander and more glorious than the dress of man, even the robes of your richest and greatest monarch. Be not therefore uneasy, for He who clothes the Lily will clothe His people too. Reflect how, though they spring from the mire and dust, the blossoms of the Lily are still pure and beautiful. Some are of the purest white, as though they had been bleached—types of innocence and purity ; others are prettily tinted with pink and red, yet they have not called in the painter to bedeck them, or asked for the services of the fuller to whiten them. Look at the texture of the leaves, and examine the cells of the petals, the woof and warp, so to speak, of the delicate dress. Yet they did not spin. The joints and curves are perfect and regular, yet they did not toil in order to reach this perfection. Examine them as you will, under the most powerful lens, and the more you study the more will you admire the perfection of form and outline, and the finish of every part. Now, since the flowers are creatures of a day, it is reasonable to think that the care which they receive is proportionate to their value and importance. We are of greater value than many Lilies, and may, therefore —if we seek Divine assistance, and place ourselves in

Almighty hands—expect that the same care will be
bestowed on us, in proportion to our value in God's sight,
as is bestowed on them.   Thus the flowers teach us the
lesson of restfulness.

I had intended, in this book, to dwell on some other
topics of interest, but am constrained to act on the advice
of Dryden, who tells a man not to write all he can, but
all he ought.   I could have written on the virtue of
purity, and this would have been a most congenial subject ; but the suggestion of the topic may perhaps suffice
to lead the reader to think of it for himself.   Perseverance
is another virtue for which many of the flowers are noted,
and this lesson we might ponder with profit.   But there
are other lines of thought to be followed, and I shall now
pass on to consider some traits and features of life, which
can neither be regarded, in the strictest sense, either as
virtues or vices, yet which go a long way towards making
variety and averting monotony, causing joy or producing
sorrow, scattering happiness or bringing about misery,
wherever they exist.   What there is in these things of
good let us imitate, what is mean and ignoble let us
avoid : so shall our own life be pure and blessed, and
that of others made happier and better.

BOOK IV.

The Ministry of Flowers

RESPECTING VARIOUS FEATURES OF LIFE,

" The love of Nature, and the scenes she draws,
Is Nature's dictate."

Cowper.

BOOK IV.

𝔙arious 𝔉eatures of 𝔏ife.

NATURE seems to mingle jest with earnest. Sometimes we see her in sombre hues, and we are made sad ; at other times she is so joyous and gay that we are immediately cheered by a sight of her face ; but there are seasons when we cannot tell exactly what she means, and we have to ask whether she is only playing tricks with us, or if she is in sober earnest. Some of these peculiarities it will now be our duty to mention, and we shall endeavour to ascertain to what extent it may be worth our while to copy the example, or avoid the influence, of these sagacious plants.

PLANT MIMICRY

Is a subject fraught with interest, but as yet it has not been investigated with sufficient thoroughness to enable

N

us to say, in every instance of supposed mimicry, which is the original, and which the imitator, what end is answered by the feigned appearance of similarity, and to what extent the art has been evolved in the history of the plant. These are questions which will yet demand the careful research of naturalists, and their study will bring its own reward to those who indulge in it. In Africa there is a species of tall red antelope, which so exactly resembles the hills thrown up by ants, that the sportsman finds himself frequently deceived by the similarity. The antelopes, " being a deep red-brown in colour, and standing one by one stock-still at the approach of the caravan, they deceived even the sharp eyes of my men (says Mr. Johnston), and again and again a hart-beest would start up at twenty yards' distance and gallop off, while I was patiently stalking an ant-hill, and crawling on my stomach through thorns and aloes, only to find the supposed antelope an irregular mass of red clay." So, too, we find caterpillars exactly resembling sticks on leaves, and butterflies simulating the tints of the foliage on which they settle, while a variety of similar curious facts are constantly thrusting themselves problematically upon the notice of the observer of nature. It is with plants and flowers, however, that we are here concerned.

Let us begin with the Orchids, a most curious and interesting group of plants, of which the Lady's Slipper

is an English representative. In this case there is
little, if any, marked resemblance to insects or other
creatures ; but in many varieties, both native and
foreign, we can at once see in the flowers the shapes
and markings of bees, butterflies, and other things.
I have specimens from Tasmania and New South
Wales, which, when growing, are so exactly like
butterflies or moths settling with outspread wings on
a strand of grass or a leaf, that my friends assure me
they have sometimes gone cautiously up to the flower and
swooped it off as they would do an insect, not knowing
until they looked at the capture but that they had secured
a bright-winged insect ! Referring to facts like these,
one writer asks : " What is a plant ? What do we mean
by the word vegetable ? It is a term the meaning of
which the most ignorant presume they understand, although the most learned are unable exactly to define it ;
for a plant is, indeed, as Theophrastus long ago observed,
'a various thing, of which it is difficult to give a definition.' Tell a clown it is difficult to distinguish an animal
from a plant ; he will smile incredulously, and perhaps
will say : 'Can I mistake Man-Orchis flowers for men ?'
but show him a Conferva and Polype, a Lichen and a
Coralline, a Flustra and a Flag, or even a Mushroom and
a Medusa, and he will at once confess, at least by silence
if not by words, that he 'kens not which they be'"

("Outlines of Botany," p. 15). Before returning to the Orchids, I will supply one other illustration of this subject which came yesterday under my notice. I had been out searching for Mosses, and had succeeded in discovering some very tiny varieties with bright red capsules modestly seated on delicate green fronds. Presently I came to an old wall on which I was certain some three or four varieties of Moss would be found, and surely enough there they were, with their fruit in most instances rising on stalks an inch or more in length. But there are cases in which the fruit is sessile, or has little if any seta. While I was looking for specimens of the varieties which were ready for the herbarium, what was my delight to see what I took to be a new species of Moss with which I had never met before. It was procumbent and branched, while here and there, prettily dotting the fronds, were little orange spots, evidently (I thought) the sessile fruit. So charmed was I with the colour of these spots, that for a few seconds I was lost in admiration, then took out a strong pocket lens to see whether the peristome was composed of a single or double row of teeth. On bringing one of the spots into focus, I found it appeared to be moving, so taking up a position which was more suitable for careful examination, I studied the spots with my strongest glass. Now, instead of tiny thecæ belonging to the Moss, I found that the bright

ORCHID. [*face p.* 196.

specks were spiders of most delicate form, almost exactly
like the little red Water-Mite (*Eylaïs*), and presenting
under the microscope the most finished and beautiful
form.   I will not call this a case of mimicry, but it
ranks well with many of the curious facts which careful
observers are daily discovering in the animal and floral
world.

In our own flora we have various representatives of
the Orchid family, and in some instances the resemblance
of the flower to certain members of the animal world is
not a little striking.   We have the Bee, Fly, Butterfly,
Spider, Frog, Monkey-Orchis, with others, and the foreign
list greatly increases the number.   While we, for example,
have a Lizard Orchid, the flora of Venezuela can claim
a Zebra Orchis (*Oncidium Zebrinum*) : so called because
its white petals bear violet transverse bars which look
much like the stripes on that pretty animal which every
visitor to the Zoological Gardens has seen.   Elsewhere
there grows a flower which has been called the Snipe
Orchis, and these are but a few out of the many plants
which belong to this large and interesting family of
mimics.   In some instances it is altogether impossible
for us, with our present knowledge of the subject, to say
what end can be answered by the adoption of these
curious forms on the part of flowers.   One can imagine,
for example, that the butterfly-shaped Orchids of Australia

would attract insects toward themselves, and so secure
the aid of their lepidopterous visitors in the work of
fertilization; but then the same does not always hold
good.   Take, for example, our Bee Orchis (*Ophrys
apifera*), which certainly looks very like the insect whose
name it bears.   In this case "there is what might be
thought a case of protective resemblance, the flower being
so fashioned as to attract bees to assist in its fertilization.
But, on the contrary, the Bee-Orchid is one of the few
plants of its order that appears to be perpetually self-
fertilized, never being visited by insects."   On this great
question no English naturalist has written more learnedly
or toiled more ardently than the late Dr. Darwin, whose
book on the Fertilization of Orchids will long remain the
standard work of reference.   It is curious to observe
what apparent freaks flowers are subject to ; for while
one kind of Orchid (*Epipactis latifolia*), which greatly
resembles the Bee-Orchis, is fertilized by wasps, its sister
(*E. viridifolia*) is independent of insect agency and fertilizes itself.   It has been suggested by some writers that
the Bee-Orchis (*Ophrys apifera*) has acquired its resemblance to that insect in order to frighten other
insects away.   This shows how little able we are as yet
to read all the mysteries of nature ; for it certainly does
not seem likely that one Orchid would acquire the shape
and appearance of a wasp in order to attract the attention

of that creature, while its sister would simulate a bee in order to scare the insect and prevent it alighting. In some instances, no doubt, the flowers may be in a transition state, and are passing from the older and less healthful habit of self-fertilization to the more recent but most valuable one of cross-fertilization. Old Gerarde, as might be expected, has a number of quaint and interesting remarks on these peculiar flowers, but we must refrain from quoting him here.

"Why, what is that, father?" asked my little girl the other evening. "Is it a goat? It looks just like one." A book was lying open on my study table, and the object which had suggested the question was the picture of the curious plant known as the Tartarian lamb (*Aspidium Barometz*), a species of Fern. A great deal has been written respecting this very curious vegetable product from the time of Struys, who travelled through Russia and Tartary in the middle of the seventeenth century, down to the present day, when Professor A. de Gubernatis discourses about it in his work on Plant Mythology. The following extract is from the Travels of the former writer, and though a wonderful story, and much perverted, is nevertheless "founded on fact." Our author says that, "On the western side of the Volga there is an elevated salt plain of vast extent, but wholly uncultivated and uninhabited. On this plain, which furnishes all the neighbouring countries with salt, grows the Boranetz or Bornitsch. This wonderful plant has the shape and appearance of a lamb, with feet, head, and tail, distinctly formed. Boranetz, in the language of Muscovy, signifies a little lamb, and a similar name is given to this Fern.* Its skin is covered with a very white down, as soft as silk. The Tartars and Muscovites esteem it highly, and preserve it with great care in their houses, where I have seen many such lambs. The sailor who gave me one of these precious plants, found it in a wood, and had its skin made into an under-waistcoat. I learned at Astrakan, from those who were best acquainted with the subject, that the Lamb grows upon a stalk about three feet high, that the part by which it is sustained is a kind of navel, and that it turns itself round, and bends downward to the herbage which serves for its food. They also said that it dries up, and pines away when the grass fails. To this I objected, that the langour and occasional withering might be natural to it, as plants are accustomed to fade at certain times. To this they replied, that they had also once thought so, but that numerous experiments proved the contrary to be the fact : such as cutting away, or by other means corrupting or destroying the grass all

---

* The linguistic student will recognize the connection between this word, and the Russian name for sheep. "Kaempfer says that the sheep of the country are called by the people dwelling on the borders of the Caspian Sea, Borannek."

around it ; after which, they assured me, it fell into a languishing state, and decayed insensibly. These persons also added, that the wolves are very fond of these vegetable Lambs, and that they devour them with avidity, because they resemble in taste the animals whose name they bear ; and that, in fact, they have bones, blood, and flesh : and hence they are called Zoophytes, that is, plant-animals. Many other things I was likewise told, which might, however, appear scarcely probable to such as have not seen them." Surely in this case the plant would not gain much by its mimicry, but if all this were true we should suppose it was a wise provision on the part of nature for rescuing the valuable sheep from the hungry wolf, by placing a vegetable substitute before it. This curious plant is found in South China, and may be procured at Canton or Macao, but, as in many other cases, the aid of the imagination must be called in to enable one to see the exact representation of the Lamb in its form. Dr. Darwin, in his " Loves of the Plants," written towards the end of the eighteenth century, says :

> " Cradled in snow, and fanned by Arctic air,
> Shines, gentle Barometz ! thy golden hair ;
> Rooted in earth, each cloven hoof descends,
> And round and round her flexile neck she bends,
> Crops the grey coral-moss, and hoary thyme,
> Or laps with rosy tongue the melting rime,
> Eyes with mute tenderness her distant dam,
> Or seems to bleat, a Vegetable Lamb."

The reader will find the subject still further discussed in the writings of such travellers as Yule, and such scholars as Scaliger, A. de Gubernatis, and others. I may here remark that in Russia the Cowslip (*Primula veris*) is called *Barancik*, lamb, or "petit agneau."

Before we leave the question of plants taking the forms of animals and insects, let us look at the fact that in the case of many fruits there are also striking resemblances. Dr. Cooke in his entertaining "Freaks and Marvels of Plant Life," gives us an illustration of the so-called Snake Nut, a fruit discovered in Demerara about half a century ago. When the nut is opened and the membrane removed, the kernel presents a striking resemblance to a snake coiled up. "There was the head, the mouth, the eyes, so complete, that one unacquainted with the fact would have believed them to be an imitation made by human hands, and not a freak of nature. As is often the case with the productions of the interior, the colonists were entirely unacquainted with the mode of growth of the plant which produced these strange nuts. . . . From the resemblance of the kernel to a snake, it was supposed that it might prove an antidote to snake poison." The reference in this last sentence to the doctrine of signatures reminds us that much may be learned respecting the similarity in the shape of leaves, flowers, and fruits, to animals and organs of the human body, by a study of the old herbalists.

In China I have frequently seen the women searching for the fruit of the Water Chestnut (*Trapa bicornis*) which, as its trivial name implies, has two horns exactly resembling the head of a bull. The people are very fond of this fruit, and I have many a time been kept awake for hours by the mongers who were hawking them in their boats up and down the river till midnight.

To turn to the curious similarity between plants of very different orders : I have no doubt that many young botanists have mistaken the pretty Hare's Ear (*Bupleurum*) for a Spurge (*Euphorbia*). For some time after I had commenced the study of botany, a specimen of the former lay in my herbarium among my various species of Spurge, it being impossible for me to identify the plant with any of the Euphorbias of my text-books ; and yet the thought never occurred to me that the specimen belonged to another family, till an accident suddenly opened my eyes and revealed the secret. Another of the umbels greatly puzzled me once, for when I first saw it growing, the flowers being as yet imperfect, I concluded from the palmate shape of the leaves that it must be one of the Hellebores, and belong to the Ranunculus order (*Ranunculaceæ*). When the flowers appeared they proved a puzzle, for they differed entirely from all the Hellebores with which 1 was familiar, and at last I ascertained that it ranked with quite another order (*Umbellifera*), and was

called *Astrantia major.* My young friends have often
been puzzled by the comparison and contrast of the
Penny-wort (*Cotyledon umbilicus*) and the marsh Penny-
wort (*Hydrocotyle vulgaris*), yet with all their similarity
they are very widely separated.

How many plants there are whose leaves are similar.
We take the Ivy as a model, and find a Speedwell, Toad-
flax, Nepeta, and other plants with leaves of a similar
shape. In like manner the Cactus has its mimic in the
Euphorbia, the Horse-tail (*Equisetum*) in the Mare's-tail
(*Hippuris*), the plants in either case being far removed in
organic structure from those of the mimic. The Rock-rose
(*Helianthemum*) which in England forms an order by itself
(*Cistaceæ*) is very similar in appearance to a Buttercup
on the one hand, and a Potentilla on the other, yet there
is no relation between them whatever.

Instances might be multiplied indefinitely, and the
reader will find it interesting to make a note of striking
similarities between plants of different orders, and the
mistakes he falls into in trying to decide to what family
a given flower belongs. The fact that plants of widely
different orders have leaves, roots, tendrils, flowers or
fruits which are similar to those of other orders, genera,
or species, makes the botanist very careful how he passes
his opinion on a critical plant unless he has the whole of
the essential organs before him. The elegum-bearing

plants have been called papilionaceous, because in many species the flowers (as in peas, beans, and broom) bear a striking resemblance to the form of a butterfly ; yet who, on seeing an Acacia in flower, would suppose that it was related to these well-known plants? Its flowers have not the insect form, but the fruit is a legume nevertheless.

The adoption of these curious forms on the part of plants must be intended for some useful purpose. In some instances the end to be attained can be at once detected, as when a flower secures insect fertilization by attracting to itself the bee or fly which its own form resembles. In other cases the problem is as yet unsolved, and the field for patient investigation is an inviting one for such as wish to know what study is worth their attention. Look at the Geraniums, for example, and observe how one has leaves like an Oak, another the form of Ivy, and yet another the leaf of an Anemone. One, whose home is in Switzerland, has been called the aconite-leaved, and another, which grows in Italy, has a tuberous root. Miss Kent remarks "that it is curious to observe how some plants appear to be compounded of others." Thus the Japanese Camellia has been noticed as resembling a Bay-tree with Roses ; the Arbutus is like another species of Bay, yielding Strawberries ; and the Laburnum seems like a tree made up of large Trefoil and garlands of yellow Pea-blossom. While one species of Milkwort

(*Polygala*) has leaves like Furze, another like Box, and a third like Myrtle, another species (*Securidaca*) has winged seeds like the Samara of our Maple.   But examples must not be multiplied, for we would again impress the fact that we wish, not so much to supply all the facts that have been collected by others, as to lead the reader to begin for himself a personal investigation of these freaks of nature.

THE ART OF WINNING.

IT is now a well established fact that the flowers adopt different devices for securing the services of insect and other visitors, by means of which self-fertilization is secured. Sometimes this end is secured by means of perfumes, sweet as attar of Roses, sometimes by means of curious bracts, leaves or appendages of various kinds around the floral organs, but most generally by means of gaudy petals, perianth, or corolla. That such an end is kept in view may be fairly argued from the fact, that those flowers which are self-fertilized grow in winter and spring when insects are rare, or depend upon the passing breeze or shower, seldom have showy flowers. Take, for example, the wind-fertilized (anemophilous) flowers, represented by such catkin-bearing trees as Willow, Hazel, Oak, Alder. The long tassels of yellow and red are pollen-bearing or staminate flowers, and the pistillate forms are usually so insignificant that only the botanist is aware of their existence and use. If a bee or other insect were to visit these tassels it would only be for the sake of the honey or pollen, for it would certainly not visit the flowers which have pistils awaiting fertilization by the catkin's pollen. Here then, while the catkins are very graceful, and make a lovely show in February and March—

> " While the trees are leafless,
> While the fields are bare,"

no end would be gained by the production of richly coloured blossoms, and the material which would be required for the perfection of such organs can now be employed in making a large supply of flour or pollen, to ensure a sufficient quantity falling on the female flowers whenever the wind blows it away from the anthers.

The same argument applies to such flowers as the Groundsel, Chickweed, Whitlow-grass, Bittercress, and some species of the Buttercup, together with many other early spring wild-flowers. If you look at them carefully you will find the following among other facts: (1) Some of them are reproduced by bulbs or corms, as the Snowdrop, (2) some by tuberous growths, as the lesser Celandine (*Ranunculus Ficaria*), (3) some by division and multiplication of the root, as the Primrose, (4) some by seed which is produced by self-fertilization. In all these instances the plants can be entirely independent of insects, consequently they do not require showy blossoms. In some cases the flowers profit by the visits of insects, although their continued reproduction would be ensured without their agency. The Primrose, for example, often profits by the kindly attention of bees, butterflies and other insects, and from the seed thus secured, new varieties may be obtained. But if you will look at some flowers you will find that it is utterly impossible for them to produce seed or propagate their species without the aid of insects, animals or man. The gardener is well aware that his Apricots, Nectarines, and other fruit trees, will be fruitless, no matter how much blossom they have, if bees are not admitted to the house, or his own hands are not busily employed in touching the pistils of the flowers with pollen-bearing stamens. Sometimes the pistil is so much longer than the stamens that the pollen of the same flower cannot possibly reach it ; at other times the stamens ripen before the pistils or *vice versa*, while yet again the anthers will open outwards and shed the pollen in such a way that it all falls away from, instead of towards, the stigmatic surface of the pistil. In all these, and a variety of other cases, as, for example, when the pistils grow on one flower or plant, and the stamens on another, insect agency is indispensable.

A fragrant smell is emitted by many flowers in order to attract the attention of those creatures whose services are required. The little leafless Daphne (*D. Mezereum*) is exceedingly sweet. There is a lilac or purple flowered variety which grows wild in some parts of England, and is frequently cultivated as well; there is also a white flowered kind which is found in gardens. The country people will tell you that the purple is the male and the white the female shrub, but you must not place too much faith in such assertions. In each case the organs are the same, but the white variety appears to me to be sweeter than the purple, and, though I have not been able to verify the matter by personal observation, I should readily believe that the white variety is visited more largely than the purple by moths and night flying insects. Here observe that, as a rule, the most fragrant flowers are not the most gaudy, nor are the gayest generally famous for their scent. The white Hyacinth, Daphne, Violet, and other modest flowers are able to convert into perfume the material which the bright scarlet Geranium or large flowered Holyhock puts into pigment. Hence it will be found that night flowering plants are usually fragrant, and have petals either white or yellow, or a slight modification of these simple colours. Pass through a field where the white Campion (*Lychnis vespertina*) is growing profusely, and while you will scarcely notice their presence during the day you will be amazed about eventide at the rich perfume which is wafted across the field, and the dazzling purity of the large open-eyed blossoms. The colour and scent are both calculated to attract moths, and if you are an entomologist, or delight in collecting butterflies and moths, this will be your harvest field. The évening Primrose (*Œnethera biennis*), the sweet Crocus, and many other plants rank with the Campion and Stock as night odorous flowers. Some have supposed that the Snowdrop should be included, but whether it enjoys the favour of insect visits I cannot say. The Tobacco plant produces blossoms which are much prized by many people who cultivate it as a window plant on account of its fragrance.

Sweetness of odour is not, however, by any means confined to night-flowering species. We may mention the Violet, Narcissus, Alyssum, Wood-ruffe, Honeysuckle, Heather, Rose, Sweet-brier, Dianthus, Orchis, wild Mignonette, and Meadow-sweet as a few representatives of our British wild flowers with a sweet smell. In some instances the scent is accompanied with a large amount of honey or other sweeter substance, while in other instances the honey exists without any remarkable perfume, so far as we can detect, but with flowers of various hues. The leguminous plants such as Clover, Peas, Gorse, Broom, and Vetch seem to be large producers of honey; so are the Labiate and Composite orders in many instances. In the Lonicera this is so remarkable that we have by universal consent called it the Honeysuckle, a name which is applied in many places to the catkins of willow or flower heads of clover, as well. Now this supply of honey was primarily intended for the benefit of the plant itself; but since the best way by means of which the flower could be benefited would be by the visits of butterflies and bees, the flowers have taken the insects into partnership, the former being in fact " sleeping partners " finding the capital by means of which business can be carried on. The harmony of nature in this respect is exceedingly interesting and beautiful. Yet even here we find law-breakers : insects which will steal the store of honey without fertilizing the plant. If you will study some of the commonest plants carefully during the spring time you will find that their secret chambers have been broken into by insects armed with saws, crowbars, and pickaxes, in the shape of organs attached to their mouths, and with these they have succeeded clandestinely in admitting the insect to the honey store which has been " with malice aforethought " taken away. In this way floral good nature, like that of man, is sometimes imposed upon, and to prevent this many flowers have armed themselves with spines, horny material or silex.

In addition to the use of fragrant odours and sweet secretions, we find also the employment of highly-coloured leaves. Sometimes these are scarlet bracts as in the now well-known Poinsettia, which makes a very showy appearance, even though its actual blossoms are insignificant.

As these highly-coloured leaves would attract day visitors, so in China I have seen large plants bearing white bracts of a similar kind, which would seem to be intended to attract night flying insects. Similar to this is the method adopted by the Guelder Rose (*Viburnum*), which has a large head of white flowers, the outer row of which consists of barren florets, which attract insects towards the plant when the central blossoms, which are fertile, lay claim to their sympathies and assistance. The large heads which are characteristic of so many of our common umbels, and such plants as the Elder, are designed to give the plants prominence, their size at once compensating for want of colour and securing the fertilization of a large number of ovules.

Of large, showy blossoms, it will scarcely be necessary to speak, for every one will recognize the immense advantages enjoyed by plants which can advertise their wares in bright colours and bold type. The flora of this country does not present to our view such astonishing plants as we meet with abroad, but even here we have flowers of which no country need feel ashamed. In the East vegetation is luxuriant, and the insects are much larger than our own. Look at the cases of butterflies and moths which come from India, China, the West Indies, and other tropical lands. Such aristocrats would think it beneath them to notice many of the small blossoms which cater so successfully for our own insects. They must have plants whose petals are large and showy, hence we find the trees, shrubs, and even humbler plants putting forth enormous blossoms, and the first glimpse one obtains of tropical vegetation is perfectly amazing. We can form some idea of the general appearance of a forest, hedgerow, or field in foreign lands by a visit to Chatsworth or Kew, and when we have looked upon the enormous Cactus flowers, Orchids, and Azaleas, not to mention plants less known to the general reader, we shall readily understand that insects and flowers in the East are as beautifully adapted to each other as they are in the West. It is scarcely true that flowers which grow in tropical lands are without fragrance. The Narcissus, Jasmine, and very many other plants, are surpassingly sweet; but at the same time, as gaudy birds are not usually great songsters, so gaudy flowers are not great perfumers. This is natural, and shows with what care Nature is adapted to meet the ends she has in view. The plumage of Australian birds is sufficient of itself to secure for the proud creatures such attention as they require without the need of song; but our modest lark, thrush, or robin, wanting in plumage, make up for the loss by pouring out strains of sweetest melody. The Scarlet Geranium is sufficiently attractive, with its head of well known flowers, without the aid of a scent-bottle; but the

ELDER. [*face p.* 214.

pale and modest Lily of the Valley might be overlooked, did it not send out upon the evening air a delicate and sweet perfume.

The colours of flowers in their relations to insects are very interesting. Persons who reside in the neighbourhood of Richmond or Oxford will be well acquainted with the Fritillary (*Fritillaria Meleagris*) with its chequered petals, somewhat resembling a chess board. Sometimes we find specimens which are entirly white, others are a compromise between the lighter and darker varieties. In some cases the existence of flowers whose colours fluctuate in this way indicates that a change is coming over their habit, and affords ground for the evolutionist's theories. I may mention the Hepaticas, Forget-me-not, and wild Geraniums. Freaks of nature, however, are common—at least we call them freaks because we are as yet ignorant of any law which regulates them; consequently we find such anomalies as white Bluebells, white Herb-Roberts, and the like, whereas the original colour is pink, red or blue. Into this interesting study we must not venture now, although the note book of a careful student may soon be filled with interesting and instructive facts drawn from personal observation.

Not less interesting is the study of the shapes of flowers in relation to insects. Look, for example, at the

Foxglove, Harebell, or Dead Nettle, and you will find that their very contour, the arrangement of the tube, stamens and pistil, and the manner in which they are suspended, all bear evidence to the fact that the art of winning has been carefully studied. But the study of this subject, if carried out to anything like its proper limits, would fill a volume. We must therefore be content with the mention of one other point, *viz.*, that which relates to fruits. It is now generally admitted that the shapes and colours of fruits have special reference to the birds, animals, and insects which partake of them as food. Our common hips and haws are favourite dainties with certain migratory birds which carry the vital portions to various parts of the country and even to other lands, and deposit them in spots suited to their future growth. The question of the colours of fruits has not yet been so carefully studied as that of flowers, and conflicting conclusions are arrived at by scholars who have given the subject their attention, but this does not alter the fact that the bright colour of such berries as grow on the Arum, Rose, Hawthorn, Yew and other plants, shrubs and trees, is specially suited to attract the attention of birds.

The moral of all this will be patent. Surely there is something to be learnt from the study of so interesting a subject. The business man may learn that clever methods of advertising will undoubtedly result in the growth of trade ; neighbours may learn that sweetness of temper and bright smiling glances will secure the interest and affection of others ; and all of us may learn that a bright, beautiful, fragrant, and fruitful life will secure the goodwill of those who are around us, and enable us to make our mark in the world.

PARASITISM.

DURING a voyage through the Indian Ocean some time ago, the steamer on which I was travelling came to a stand. While we were waiting for it to resume its course, a cry was raised announcing that a shark was at the stern of the large vessel, seeking its breakfast. A bait of pork was provided for it, and the monster caught and hauled on board. When we came to examine it we found on its back a fish which secured its living by preying on the shark. It was simply a parasite. but of rather a large order, and, no doubt, like other fish it in turn proved host to other parasites. There is scarcely a representative of the animal world which has not one or more unbidden guests. Such unsavoury creatures as bugs, fleas, and lice are to be found not only on the persons and property of uncleanly people, but also on monkeys and bears, cats and dogs, poultry and birds, in fact, in one form or another, they abound everywhere. What is true of the animal world is equally true of the vegetable. Some parasites are exceedingly large, the Mistletoe being perhaps the best known English example. As I drive through Sherwood Forest from time to time I can see, not only on ancient Whitethorn trees, but high up among the branches of massive Poplar and other trees, the dark green boughs of this curious plant. There is a wonderful range between this parasite and that which, in the form of rust or blight, attacks and destroys our corn. The one is so large that in winter its deep coloured leaves may be detected among the bare trees for a mile; the other so small that a powerful lens is required to bring out its form and definition. Yet strange to say, these parasites have their parasites also. Upon the Mistletoe bough will be found such things as lichens and insects, and I have just found a tiny insect living on the micro-fungi of the Blackberry, and the cluster-cups (*Æcidium*) of the Berberry. In the former case, the fungi had been mounted for the microscope and examined more than once before the tiny parasite was detected among the dark brown spores; in the latter case, the little green insect tottered over the cups on the underside of the leaf, just as though it were trying to span the crater of a miniature volcano, and ever and anon it would go tumbling into the fiery lava, but came out again unscathed! Parasites upon parasites—we have but just begun to realize how everything forms a pastur-

age for something else.   Here is a tiny sprig of moss
which I found yesterday parasitic on a bush.   It is no
larger than a pin, but if you look at it you will find it has
a bright red head.   That spot, which looks like a tiny
carbuncle set on the end of a silver pin of delicate work-
manship, is a perfect fungus, as you may easily prove to
yourself by putting it under an inch lens.   Thus the
moss which preys upon another plant is in its turn the
host of a third, and each of these belongs to as widely
different orders as do the flea and the pigeon.

In the vegetable world then there is scarcely a plant or
flower which has not its parasite.   I will not here dwell
on the mosses and lichens again, as we have already seen
what is their position and purpose in the world.   To the
Dodder and some other plants attention has also been
directed in the foregoing pages, and I shall, therefore,
limit myself to a few further notes on the simplest and
minutest forms of plant parasites.   These may be found
all the year round.   The common Groundsel (*Senecio
vulgaris*), especially when growing on somewhat poor
soil, will be found from January to December, affording
daily sustenance to a minute golden fungus (*Trichobasis*
or *Coleosporium senecionis*).   I have just returned from a
stroll across a stubble field, around the headlands of
which a good many Groundsel plants are to be found.
As I pulled up one after another of these weeds I found

one, two, or more leaves bearing on their undersurface the object for which I was searching. In its early stages it is simply a smooth, shining, yellow spot or ser es of spots, no larger than a pin's head. By-and-bye, the skin bursts and a number of very minute globules are seen to be thrusting forth their heads. Later on these will have escaped, and may be seen scattered over the leaf in the form of gold dust. Their structure is very simple. If you scrape a little of the dust on to a slide, and examine it under the microscope, you will find that it consists of spores of irregular form, without appendages of any kind. In this they differ from the fungi which are found on the leaves of the Bramble and Raspberry, which are of a much more complex and interesting character.

As the groundsel rust may be found early in the year, its study may occupy us till the leaves of the Buttercup begin to appear, when we shall be able to find new objects of interest. On the leaves of the lesser Celandine (*Ranunculus Ficaria*) we shall find, in early spring, speci· mens of an interesting Clustercup, and when these make their appearance the harvest of the micro-fungi hunter has commenced. These Clustercups are delightful studies as opaque objects under a low power, and most of our readers will prefer to see them in their simple state, leaving to the specialist the task of mounting sections and examining the parasites minutely. When once found on

the under side of a leaf their form will readily be recognised again, for they have exactly the form and
appearance of fairy cheese-cakes, and look very much like
certain forms of Pezizæ or the so-called Fairy Baths.
You will now find them abundantly on the leaves of the
troublesome Coltsfoot, on the Berberry, Primrose, Daisy,
Dandelion, Violet, Geranium, Bedstraw and Nettle, as
well as on' other plants.   In some instances you will find
the leaves of a plant attacked with two or three species
of fungi at the same time.

Here is a leaf of the wood Anemone.   It has not
only a large number of Clustercups contorting and
modifying its segments, but there is a Puccinia abounding
on it as well ; and this leaf of the Willow-herb (*Epilobium*)
is in the same predicament.   Whether you go into the
meadow or copse, the garden or orchard, the cornfield or
lane, everywhere, when your eyes are practised, you will
find these interesting objects.   Several kinds may be
found even in the depth of winter.   Not only does the
green mould (*Peronospora*) now abound, but the leaves of
the Bramble will be found affording nutriment to thousands of delicate and complex forms which well deserve
your careful study.   I have spent much time in the study
of the Bramble-brand (*Aregma bulbosum*), and every time I
take it up it seems to possess a new charm.   Those who
have microscopes, and can manipulate the slides, should

watch the breaking up of spores by the action of warm nitric acid, for the experiment is most simple and delightful. Put a cluster of spores on your slide with a drop of strong acid, cover with a thin glass circle and heat gently over the lamp. When the air bubbles begin to appear, slip the glass under your lens and watch the result. You will want all your friends to come and witness it.

Even in winter you will find some of the grasses infested with fungi, and last winter I came across a most interesting species of Ustilago in a wood, where too I found the skeletonized capsules of the wild Hyacinth and its seeds covered with another interesting parasite. But summer and autumn are the seasons in which these minute plants abound and may be found in perfection. The pear trees in your orchard will be found to afford types worth investigation (*Ræstelia*), while the potato, onion, parsnip and lettuce are but a few of the garden crops which are subject to attack and often suffer exceedingly. Even the herbs are not exempt, for the Mint and Sage, as well as the dead stems of asparagus, are frequently found supporting a host of these unwelcome parasites. It is only here and there that we find the Box-tree growing luxuriantly in a wild state, but I have recently found this evergreen attaining the size of moderate trees and bringing forth fruit. Examining the seed vessels of a number of these plants the other day, I found that they too were attacked by a fungus of very tiny growth, which was nevertheless performing its work of decomposition with considerable rapidity and success.

Fungi are not the only parasites which prey on plants, but this family is so large and varied that volumes have been already filled with their life-history, and when we remember that some of them, such as Ergot, are of use medicinally, that others create fermentation, as the yeast plant, others cover our bread and cheese with mould, and yet others destroy our vegetables and cereals, we shall see that they are worthy our study and demand attention. Parasites are not usually genial or welcome companions, yet they have their uses. The flea is anything but a pleasant associate, but its presence indicates that the broom, soap, fresh air, and other conducives to health are needed, and if its warning voice be heeded disease may often be averted. The parasites on leaves prove destructive to many plants, but by their agency the law of circularity in life is kept a-going, and if you were to study the spores of moulds and brands, which are in some instances so minute that thousands would be required to fill an inch of space, you will see what a great deal must be done by their agency to reduce larger growths to dust and powder and so create new soil for nobler plants. There is a good deal of parasitism in human society also which is equally useful. Mansions and manorial halls are like the large plants; servants, tradesmen, and beggars of various kinds are the parasites which keep the great mass from accumulating too much life and strength. They break up the gold into smaller portions, and set it free for flowing into new channels, and stimulating new enterprises, and so the world continues to go round, and life to see its needed changes and revolutions.

SENSITIVENESS AND IRRITABILITY.

ONCE spent some time at Penang, an important and lovely spot in the Straits of Malacca. As it was the day of the March full moon, a most important ceremony was being observed by a certain religious sect which is largely represented there, and everybody was going to the Waterfall, near which a shrine in honour of one of the gods of the Hindû Pantheon was erected. It was a lovely day, and a party of us started off to spend a few hours on the hills, visit the stream and temple, and see the sights of the occasion. At the foot of the hills we left our Gharî, or conveyance, and proceeded to finish the journey on foot. Groups of people were to be seen in every direction, some begging, some selling curiosities, some bathing, some feeding—all happy and good tempered. Presently, as we were climbing a gentle ascent our faces bent earthwards, what was our surprise at

finding that all around us the vegetation for two or three
feet seemed to be alive with motion. Every time we set
a fresh footstep, scores of leaves suddenly collapsed and
quivered as though they had been affrighted animals.
A glance was sufficient to reveal the fact that we were
walking on the enchanted carpet of a Mimosa bed, and
we realised more vividly than ever before the accuracy of
the name, " Sensitive Plant." Many of my readers have
already made themselves acquainted with the fact that
plants possessed of sensitive qualities are not confined to the Mimosa family or to the East ; but it may
not be so familiar a fact to others that our own flora
supplies us with a number of instances of plants which
are constructed with such delicacy that they recoil from
the rude touch of Nature's finger, close as her chilling
breath passes over them, twist themselves like worms
before the gardener when a change of weather is pending,
hide their face in a storm, and close their leaves or petals
as the shades of evening approach, opening them again
with the greatest precision as the morning dew reflects
the image of the orb of day. Others, like the well-
known and curious Sundew (*Drosera*) and Venus' Fly-
trap (*Dionœa*), are sensitive to the touch of animals and
insects, and their study is of the profoundest interest.
Let us begin with plants which may be met with most
commonly. And first on the list we will place the so-

called Hygrometric Moss (*Funaria hygrometrica*), the peristome and fruit-stalk of which act like a barometer, by the way in which they indicate very slight variations in the moisture of the atmosphere. In this Moss, as in many others, the fruit consists of a kind of urn or capsule containing spores, which are kept from escaping by means of a single or double row of teeth. Not only does the fruit-stalk twist up like a cord when the weather changes, but the teeth contract and close the mouth of the urn in wet weather, so that the contents of the theca can only be dispersed under favourable atmospheric conditions. Thus the sensibility of the plant is made subservient to a great end, and helps to ensure the proper dispersion of spores on whose germination the continuance of the race depends.

In former times, the superstitious were frequently deceived by designing men by means of awns of the wild Oat, which were said to be the legs of an Arabian spider. These awns, when damped or moistened either by the breath or by a drop of water, are subject to strange contortions and wrigglings, and these movements were interpreted by the jugglers in such a way as to delude the ignorant. Now, the possession of such appendages on the part of seeds is often of great value, for by their means the seed is enabled to fix itself in the soil, where it is held until roots are thrown out and germination commences. The elaters which are found in the Horse-tails (*Equisetum*) are useful for the same purpose, seeing that they not only serve to disperse the spores, but by their sensitiveness to moisture readily open on touching the damp surface of the earth, and so clutch the soil and secure a resting-place for their charge, which in due time begins to grow. The study of these elaters, which are found also in the Puff-ball, Liverwort, and other plants, is full of interest. Those who are fond of observing the freaks of Nature will find many similar instances of sensitiveness and irritability. Take, for example, the old-fashioned Balsam, which has merited the name of Touch-me-not, through its irritability. You no sooner touch one of the seed vessels which is ripening, than, with a jerk that quite startles you, it splits and curls up, throwing the seeds to a great distance. A similar act may be observed in the wild Geraniums, and if you should be preparing specimens for the herbarium which have seed vessels nearly or quite ripe, do not be surprised if you suddenly hear a crack, and see the seeds flying up to the ceiling, or across to the extremities of the room.

You will find that the flowers of the Dog's Mercury (*Mercurialis perenne*) are often subject to fits of irritability, and by this means the staminate flowers are sometimes enabled to cast their pollen on to the viscid surfaces of the pistillate blossoms. In the Berberry, Pellitory, Nettle, and a number of other plants, the stamens or pistils manifest a similar irritability.

The Corline Thistle is frequently employed on the Continent just in the same way as country folk in England use a piece of Sea-weed for indicating coming changes in the weather, and there is not a boy or girl who lives in the country but knows how regularly the little Shepherd's Weather-glass (*Anagallis arvensis*), or Scarlet Pimpernel, will close its fair petals before a coming storm. True, the little plant must not be so implicitly trusted as some people believe ; yet, as a rule, it will be very quick in its response to atmospheric warnings. The same may also be said of many other plants, which not only have regular hours for closing and opening their eyes, but will keep up their shutters on a rainy day, and open them cautiously when the weather is unsettled. Then the sleep of plants, as it is called, is another illustration of sensitiveness. Some plants are regular sluggards, for they only open their doors at noon and close them again directly after dinner ; others rise late and retire early, while most open their petals in the morning and close them again in the evening, thus showing that they are capable of distinguishing between the various portions of the day. Some open only in the evening or early morning, their habits being as regula as those of their neighbours. Thus the Red Campion flowers by day, but the white variety blossoms at night. The exceptions with which one frequently meets would seem to indicate that the night-flowering species had not yet fully adopted the habit, for in passing through a field of corn or clover you will occasionally find a White Lychnis in bloom in the morning or afternoon. But in these cases, I have generally observed that the fragrance of the night-blooming plants has not been present, or at any rate, only to a very small degree.

But not only do flowers go to sleep, we find that the leaves of plants also assume different positions during the night from those which they maintain during the day. Plants with trifoliate and pinnate leaves, especially such as belong to the great family of legumes, represented by the clover, wood-sorrel, sensitive plant and acacias, are specially noteworthy in this respect. Writing on this subject, Dr. Brown remarks, that " Perhaps in no plants is this irritability better shewn than in some Leguminosæ, especially in Mimosas or Sensitive Plants. These plants have bipinnate leaves, with four secondary petioles starting from a common rachis (leaf-stalk) or petiole, each of the petioles being provided with a number of pairs of leaflets, which are expanded horizontally during day-light. If the common Sensitive Plant (*Mimosa pudica*) is suddenly jarred or touched, the leaflets will change

their position, overlapping one another from below upwards, close to the secondary petiole ; on greater irritation
being applied, the secondary petioles also bend forward
and approach one another, and finally the general petiole
sinks down by means of a bending at its articulation or
junction with the stem." In confirmation of what I have
said at the commencement of this section, we may quote
a few other words from this same writer's " Manual of
Botany." After stating that in our own hot-houses these
plants are rarely so sensitive as in their native climate, Dr.
Brown adds that "there the concussion caused by a horse
galloping along the road on the sides of which the plant
grows will often have the effect of causing the leaves to
fold up. We have often noticed this effect produced
by the passing of a train along the Panama railroad in
New Granada, on the sides of which the plant grows
abundantly. So sensitive are they that on one plant folding its leaflet, the contact will irritate its neighbour, and
so on—the irritability travelling along the patch almost
as fast as the traveller can keep up with it in walking."
What we thus find in this class of plants when acted upon
by an irritant in the day-time, we find to a greater or less
degree in many other plants as night approaches. The
power of folding and opening its leaves will be found
to exist, for example, in our common Clover and
Wood-sorrel. Either of these plants may be readily

watched by persons resident in the country, while the latter is capable of being cultivated with ease in a flower-pot by those who live in towns, and we can suggest no pleasanter pastime than that of studying the sleeping and waking of the leaves, together with the two kinds of flowers, the adventitious buds produced by decaying leaves and other remarkable phenomena relating to the history of the Wood-sorrel.

Acacias are now largely grown in this country, and they are among the most remarkable of leaf-sleeping plants, being, indeed, closely related to the Mimosa and Clover. But those who keep their eyes open as they take their morning and evening walks will be continually struck by the change which comes over a variety of plants with which they meet; and the greenhouse, garden, and lane will have many a lesson for those of us who are willing to learn them. Many text-books of botany now have a chapter on the subject of the irritability of plants, and those who wish thoroughly to investigate the subject will find Darwin's works on the Movements of Plants, Insectivorous Plants, and kindred topics, full of fact and romance. Perhaps the most interesting and popular digest of the subject is that by Dr. Cooke in his "Freaks and Marvels of Plant Life."

While discussing this branch of our study, it would be unpardonable were we to omit all reference to those plants

which are now commonly known as insectivorous—I
mean the Sundews and their allies. It would be impossible for me to add anything new to the facts already accumulated ; but, at the same time, every lover of plants
and student of their phenomena must be aware that these
flowers are among the most remarkable that have ever been
discovered. The term Sundew has reference to the appearance of the leaves of Drosera. Each leaf is supplied
with hairs or tentacles, on the ends of which a tiny dew-
drop appears to be lodged. When an insect alights upon
the leaf, these dewdrops act as a kind of sticky gum and
prevent its escape, the leaf meanwhile closing on its prey
and strangling or stifling it to death. Both in America
and on the Continent, as well as in our own country,
careful investigations into these matters have been pro-
secuted, with the most interesting and startling results.

In botanic gardens and conservatories, one will often
see specimens of the curious Venus' Flytrap (*Dionæa*),
with leaves notched in such a way that when they fold
up their effect is similar to that of a pair of toothed
clams, such as were so largely used a few years ago for
rats, mice, and vermin in woods. The leaves of these
plants are more sensitive even than those of the Sundew, and when an insect alights upon them it is instantly
secured. " The Drosera captures its prey by means of
its viscid secretion, and the pressure caused by the

struggles of the insect produces inflection, which it accomplishes quite at its leisure. Dionæa, having no secretion, secures its prey by instantly closing upon it. The filaments, therefore, which cause the lobes to close instantly at the lightest touch, are comparatively indifferent to prolonged pressure. These filaments have nothing whatever to do with the digestive process; they are merely the sentinels on guard to signal the approach of a victim."

These are but a few illustrations of the sensitiveness of plants and flowers, and the irritability which they manifest under provocation. How far useful ends are answered in these curious operations it is impossible in many instances as yet to say. We can at present see no more justification for the act of murder on the part of Dionæa when touched by a fly than on the part of a man when irritated by a fellow creature. The plant does not require the insect for food, at least so far as we can ascertain, otherwise it would be analogous to the slaughter of a sheep or pig by the butcher. On the other hand, a due amount of sensitiveness is most useful to such plants as Mimosa, Acacia, Clover and Wood-sorrel. And a sensitive conscience is equally beneficial to us. He who can tell when evil is approaching and will listen to the dictates of a voice within will be saved from many of the calamities which overtake others.

In concluding this brief study of some of the more re-

markable phenomena of plant life, we may observe that it
has not been our purpose to moralise too much, it being
hoped that the hints will be sufficient to suggest a line of
thought which will prove profitable to the reader, and
assist him in the pleasing task of explaining and enforcing
homely truths in a simple manner when dealing with
others. Apart, too, from the lessons which are taught us
by the flowers, and their higher ministry as expositors of
Divine love and wisdom, their study is of value for scientific, medicinal and physical reasons. Botany now ranks
high as a science, the flowers and plants supply us with
many of our most valuable medicines, and the search for
these stars of earth is one of the most enjoyable and healthful of exercises. From no other pursuit has the author
derived so much physical and mental benefit, and he
therefore confidently commends it to others.

THE END.

www.ingramcontent.com/pod-product-compliance
Lightning Source LLC
Chambersburg PA
CBHW021520210326
41599CB00012B/1319